Mathematics: A Very Short Introduction

D1342509

VERY SHORT INTRODUCTIONS are for anyone wanting a stimulating and accessible way in to a new subject. They are written by experts, and have been published in 25 languages worldwide.

The series began in 1995, and now represents a wide variety of topics in history, philosophy, religion, science, and the humanities. Over the next few years it will grow to a library of around 200 volumes – a Very Short Introduction to everything from ancient Egypt and Indian philosophy to conceptual art and cosmology.

Very Short Introductions available now:

Available soon:

For more information visit our web site

www.oup.co.uk/vsi

Timothy Gowers

MATHEMATICS

A Very Short Introduction

OXFORD
UNIVERSITY PRESS

Great Clarendon Street, Oxford OX2 6DP

Oxford University Press is a department of the University of Oxford.
It furthers the University's objective of excellence in research, scholarship,
and education by publishing worldwide in

Oxford New York

Auckland Bangkok Buenos Aires Cape Town Chennai
Dar es Salaam Delhi Hong Kong Istanbul Karachi Kolkata
Kuala Lumpur Madrid Melbourne Mexico City Mumbai Nairobi
São Paulo Shanghai Singapore Taipei Tokyo Toronto

with an associated company in Berlin

Oxford is a registered trade mark of Oxford University Press
in the UK and in certain other countries

Published in the United States
by Oxford University Press Inc., New York

British Library Cataloguing in Publication Data

Data available

Library of Congress Cataloging in Publication Data

Data available

ISBN 0–19–285361–9

1 3 5 7 9 10 8 6 4 2

Typeset by RefineCatch Ltd, Bungay, Suffolk
Printed in Spain by Book Print S. L., Barcelona

Contents

Preface

Early in the 20th century, the great mathematician David Hilbert noticed that a number of important mathematical arguments were structurally similar. In fact, he realized that at an appropriate level of generality they could be regarded as the same. This observation, and others like it, gave rise to a new branch of mathematics, and one of its central concepts was named after Hilbert. The notion of a Hilbert space sheds light on so much of modern mathematics, from number theory to quantum mechanics, that if you do not know at least the rudiments of Hilbert space theory then you cannot claim to be a well-educated mathematician.

What, then, is a Hilbert space? In a typical university mathematics course it is defined as a complete inner-product space. Students attending such a course are expected to know, from previous courses, that an inner-product space is a vector space equipped with an inner product, and that a space is complete if every Cauchy sequence in it converges. Of course, for those definitions to make sense, the students also need to know the definitions of vector space, inner product, Cauchy sequence and convergence. To give just one of them (not the longest): a Cauchy sequence is a sequence $x_1, x_2, x_3, \ldots$ such that for every positive number ϵ there exists an integer N such that for any two integers p and q greater than N the distance from x_p to x_q is at most ϵ.

In short, to have any hope of understanding what a Hilbert space is, you must learn and digest a whole hierarchy of lower-level concepts first. Not surprisingly, this takes time and effort. Since the same is true of many of the most important mathematical ideas, there is a severe limit to what can be achieved by any book that attempts to offer an accessible introduction to mathematics, especially if it is to be very short.

Instead of trying to find a clever way round this difficulty, I have focused on a different barrier to mathematical communication. This one, which is more philosophical than technical, separates those who are happy with notions such as infinity, the square root of minus one, the twenty-sixth dimension, and curved space from those who find them disturbingly paradoxical. It is possible to become comfortable with these ideas without immersing oneself in technicalities, and I shall try to show how.

If this book can be said to have a message, it is that one should learn to think abstractly, because by doing so many philosophical difficulties simply disappear. I explain in detail what I mean by the abstract method in Chapter 2. Chapter 1 concerns a more familiar, and related, kind of abstraction: the process of distilling the essential features from a real-world problem, and thereby turning it into a mathematical one. These two chapters, and Chapter 3, in which I discuss what is meant by a rigorous proof, are about mathematics in general.

Thereafter, I discuss more specific topics. The last chapter is more about mathematicians than about mathematics and is therefore somewhat different in character from the others. I recommend reading Chapter 2 before the later ones, but apart from that the book is arranged as unhierarchically as possible: I shall not assume, towards the end of the book, that the reader has understood and remembered everything that comes earlier.

Very little prior knowledge is needed to read this book – a British GCSE course or its equivalent should be enough – but I do presuppose some interest on the part of the reader rather than trying to drum it up myself.

For this reason I have done without anecdotes, cartoons, exclamation marks, jokey chapter titles, or pictures of the Mandelbrot set. I have also avoided topics such as chaos theory and Godel's theorem, which have a hold on the public imagination out of proportion to their impact on current mathematical research, and which are in any case well treated in many other books. Instead, I have taken more mundane topics and discussed them in detail in order to show how they can be understood in a more sophisticated way. In other words, I have aimed for depth rather than breadth, and have tried to convey the appeal of mainstream mathematics by letting it speak for itself.

I would like to thank the Clay Mathematics Institute and Princeton University for their support and hospitality during part of the writing of the book. I am very grateful to Gilbert Adair, Rebecca Douglas, Emily Gowers, Patrick Gowers, Joshua Katz, and Edmund Thomas for reading earlier drafts. Though they are too intelligent and well informed to count as general readers, it is reassuring to know that what I have written is comprehensible to at least some non-mathematicians. Their comments have resulted in many improvements. To Emily I dedicate this book, in the hope that it will give her a small idea of what it is I do all day.

List of diagrams

The publisher and the author apologize for any errors or omissions in the above list. If contacted they will be pleased to rectify these at the earliest opportunity.

Chapter 1
Models

How to throw a stone

Suppose that you are standing on level ground on a calm day, and have in your hand a stone which you would like to throw as far as possible. Given how hard you can throw, the most important decision you must make is the angle at which the stone leaves your hand. If this angle is too flat, then although the stone will have a large horizontal speed it will land quite soon and will therefore not have a chance to travel very far. If on the other hand you throw the stone too high, then it will stay in the air for a long time but without covering much ground in the process. Clearly some sort of compromise is needed.

The best compromise, which can be worked out using a combination of Newtonian physics and some elementary calculus, turns out to be as neat as one could hope for under the circumstances: the direction of the stone as it leaves your hand should be upwards at an angle of 45 degrees to the horizontal. The same calculations show that the stone will trace out a parabolic curve as it flies through the air, and they tell you how fast it will be travelling at any given moment after it leaves your hand.

It seems, therefore, that a combination of science and mathematics enables one to predict the entire behaviour of the stone from the moment it is launched until the moment it lands. However, it does so only if one is prepared to make a number of simplifying assumptions, the main one being that the only force acting on the stone is the earth's gravity and that this force has the same magnitude and direction everywhere. That is not true, though, because it fails to take into account air resistance, the rotation of the earth, a small gravitational influence from the moon, the fact that the earth's gravitational field is weaker the higher you are, and the gradually changing direction of 'vertically downwards' as you move from one part of the earth's surface to another. Even if you accept the calculations, the recommendation of 45 degrees is based on another implicit assumption, namely that the speed of the stone as it leaves your hand does not depend on its direction. Again, this is untrue: one can throw a stone harder when the angle is flatter.

In the light of these objections, some of which are clearly more serious than others, what attitude should one take to the calculations and the predictions that follow from them? One approach would be to take as many of the objections into account as possible. However, a much more sensible policy is the exact opposite: decide what level of accuracy you need, and then try to achieve it as simply as possible. If you know from experience that a simplifying assumption will have only a small effect on the answer, then you should make that assumption.

For example, the effect of air resistance on the stone will be fairly small because the stone is small, hard, and reasonably dense. There is not much point in complicating the calculations by taking air resistance into account when there is likely to be a significant error in the angle at which one ends up throwing the stone anyway. If you want to take it into account, then for most purposes the following rule of thumb is good enough: the greater the air resistance, the flatter you should make your angle to compensate for it.

What is a mathematical model?

When one examines the solution to a physical problem, it is often, though not always, possible to draw a clear distinction between the contributions made by science and those made by mathematics. Scientists devise a theory, based partly on the results of observations and experiments, and partly on more general considerations such as simplicity and explanatory power. Mathematicians, or scientists doing mathematics, then investigate the purely logical consequences of the theory. Sometimes these are the results of routine calculations that predict exactly the sorts of phenomena the theory was designed to explain, but occasionally the predictions of a theory can be quite unexpected. If these are later confirmed by experiment, then one has impressive evidence in favour of the theory.

The notion of confirming a scientific prediction is, however, somewhat problematic, because of the need for simplifications of the kind I have been discussing. To take another example, Newton's laws of motion and gravity imply that if you drop two objects from the same height then they will hit the ground (if it is level) at the same time. This phenomenon, first pointed out by Galileo, is somewhat counter-intuitive. In fact, it is worse than counter-intuitive: if you try it for yourself, with, say, a golf ball and a table-tennis ball, you will find that the golf ball lands first. So in what sense was Galileo correct?

It is, of course, because of air resistance that we do not regard this little experiment as a refutation of Galileo's theory: experience shows that the theory works well when air resistance is small. If you find it too convenient to let air resistance come to the rescue every time the predictions of Newtonian mechanics are mistaken, then your faith in science, and your admiration for Galileo, will be restored if you get the chance to watch a feather fall in a vacuum – it really does just drop as a stone would.

Nevertheless, because scientific observations are never completely direct and conclusive, we need a better way to describe the relationship between science and mathematics. Mathematicians do not apply scientific theories directly to the world but rather to *models*. A model in this sense can be thought of as an imaginary, simplified version of the part of the world being studied, one in which exact calculations are possible. In the case of the stone, the relationship between the world and the model is something like the relationship between Figures 1 and 2.

1. **A ball in flight I**

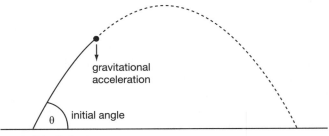

2. **A ball in flight II**

There are many ways of modelling a given physical situation, and we must use a mixture of experience and further theoretical considerations to decide what a given model is likely to teach us about the world itself. When choosing a model, one priority is to make its behaviour correspond closely to the actual, observed behaviour of the world. However, other factors, such as simplicity and mathematical elegance, can often be more important. Indeed, there are very useful models with almost no resemblance to the world at all, as some of my examples will illustrate.

Rolling a pair of dice

If I roll a pair of dice and want to know how they will behave, then experience tells me that there are certain questions it is unrealistic to ask. For example, nobody could be expected to tell me the outcome of a given roll in advance, even if they had expensive technology at their disposal and the dice were to be rolled by a machine. By contrast, questions of a probabilistic nature, such as, 'How likely is it that the numbers on the dice will add up to seven?' can often be answered, and the answers may be useful if, for example, I am playing backgammon for money. For the second sort of question, one can model the situation very simply by representing a roll of the dice as a random choice of one of the following thirty-six pairs of numbers.

$$(1, 1) \quad (1, 2) \quad (1, 3) \quad (1, 4) \quad (1, 5) \quad (1, 6)$$
$$(2, 1) \quad (2, 2) \quad (2, 3) \quad (2, 4) \quad (2, 5) \quad (2, 6)$$
$$(3, 1) \quad (3, 2) \quad (3, 3) \quad (3, 4) \quad (3, 5) \quad (3, 6)$$
$$(4, 1) \quad (4, 2) \quad (4, 3) \quad (4, 4) \quad (4, 5) \quad (4, 6)$$
$$(5, 1) \quad (5, 2) \quad (5, 3) \quad (5, 4) \quad (5, 5) \quad (5, 6)$$
$$(6, 1) \quad (6, 2) \quad (6, 3) \quad (6, 4) \quad (6, 5) \quad (6, 6)$$

The first number in each pair represents the number showing on the first die, and the second the number on the second. Since exactly six of the pairs consist of two numbers that add up to seven, the chances of rolling a seven are six in thirty-six, or one in six.

One might object to this model on the grounds that the dice, when rolled, are obeying Newton's laws, at least to a very high degree of precision, so the way they land is anything but random: indeed, it could in principle be calculated. However, the phrase 'in principle' is being overworked here, since the calculations would be extraordinarily complicated, and would need to be based on more precise information about the shape, composition, initial velocities, and rotations of the dice than could ever be measured in practice. Because of this, there is no advantage whatsoever in using some more complicated deterministic model.

Predicting population growth

The 'softer' sciences, such as biology and economics, are full of mathematical models that are vastly simpler than the phenomena they represent, or even deliberately inaccurate in certain ways, but nevertheless useful and illuminating. To take a biological example of great economic importance, let us imagine that we wish to predict the population of a country in 20 years' time. One very simple model we might use represents the entire country as a pair of numbers $(t, P(t))$. Here, t represents the time and $P(t)$ stands for the size of the population at time t. In addition, we have two numbers, b and d, to represent birth and death rates. These are defined to be the number of births and deaths per year, as a proportion of the population.

Suppose we know that the population at the beginning of the year 2002 is P. According to the model just defined, the number of births and deaths during the year will be bP and dP respectively, so the population at the beginning of 2003 will be $P + bP - dP = (1 + b - d)P$. This argument works for any year, so we have the formula $P(n + 1) = (1 + b - d)P(n)$, meaning that the population at the beginning of year $n + 1$ is $(1 + b - d)$ times the population at the beginning of year n. In other words, each year the population multiplies by $(1 + b - d)$. It follows that in 20 years it multiplies by $(1 + b - d)^{20}$, which gives an answer to our original question.

Even this basic model is good enough to persuade us that if the birth rate is significantly higher than the death rate, then the population will grow extremely rapidly. However, it is also unrealistic in ways that can make its predictions very inaccurate. For example, the assumption that birth and death rates will remain the same for 20 years is not very plausible, since in the past they have often been affected by social changes and political events such as improvements in medicine, new diseases, increases in the average age at which women start to have children, tax incentives, and occasional large-scale wars. Another reason to expect birth and death rates to vary over time is that the ages of people in the country may be distributed rather unevenly. For example, if there has been a baby boom 15 years earlier, then there is some reason to expect the birth rate to rise in 10 to 15 years' time.

It is therefore tempting to complicate the model by introducing other factors. One could have birth and death rates $b(t)$ and $d(t)$ that varied over time. Instead of a single number $P(t)$ representing the size of the population, one might also like to know how many people there are in various age groups. It would also be helpful to know as much as possible about social attitudes and behaviour in these age groups in order to predict what future birth and death rates are likely to be. Obtaining this sort of statistical information is expensive and difficult, but the information obtained can greatly improve the accuracy of one's predictions. For this reason, no single model stands out as better than all others. As for social and political changes, it is impossible to say with any certainty what they will be. Therefore the most that one can reasonably ask of any model is predictions of a conditional kind: that is, ones that tell us what the effects of social and political changes will be if they happen.

The behaviour of gases

According to the kinetic theory of gases, introduced by Daniel Bernoulli in 1738 and developed by Maxwell, Boltzmann, and others in the second half of the 19th century, a gas is made up of moving molecules, and many of its properties, such as temperature and pressure, are statistical properties of those molecules. Temperature, for example, corresponds to their average speed.

With this idea in mind, let us try to devise a model of a gas contained in a cubical box. The box should of course be represented by a cube (that is, a mathematical rather than physical one), and since the molecules are very small it is natural to represent them by points in the cube. These points are supposed to move, so we must decide on the rules that govern how they move. At this point we have to make some choices.

If there were just one molecule in the box, then there would be an obvious rule: it travels at constant speed, and bounces off the walls of the box when it hits them. The simplest conceivable way to generalize this model is then to take N molecules, where N is some large number, and assume that they all behave this way, with absolutely no interaction between them. In order to get the N-molecule model started, we have to choose initial positions and velocities for the molecules, or rather the points representing them. A good way of doing this is to make the choice randomly, since we would expect that at any given time the molecules in a real gas would be spread out and moving in many directions.

It is not hard to say what is meant by a random point in the cube, or a random direction, but it is less clear how to choose a speed randomly, since speed can take any value from 0 to infinity. To avoid this difficulty, let us make the physically implausible assumption that all the molecules are moving at the same speed, and that it is only the initial positions and directions that are chosen randomly. A

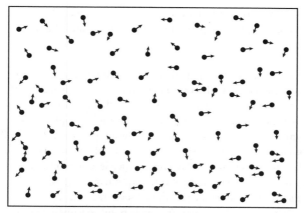

3. A two-dimensional model of a gas

two-dimensional version of the resulting model is illustrated in Figure 3.

The assumption that our N molecules move entirely independently of one another is quite definitely an oversimplification. For example, it means that there is no hope of using this model to understand why a gas becomes a liquid at sufficiently low temperatures: if you slow down the points in the model you get the same model, but running more slowly. Nevertheless, it does explain much of the behaviour of real gases. For example, imagine what would happen if we were gradually to shrink the box. The molecules would continue to move at the same speed, but now, because the box was smaller, they would hit the walls more often and there would be less wall to hit. For these two reasons, the number of collisions per second in any given area of wall would be greater. These collisions account for the pressure that a gas exerts, so we can conclude that if you squeeze a gas into a smaller volume, then its pressure is likely to increase – as is confirmed by observation. A similar argument explains why, if you increase the temperature of a gas without increasing its volume, its pressure also increases.

And it is not too hard to work out what the numerical relationships between pressure, temperature, and volume should be.

The above model is roughly that of Bernoulli. One of Maxwell's achievements was to discover an elegant theoretical argument that solves the problem of how to choose the initial speeds more realistically. To understand this, let us begin by dropping our assumption that the molecules do not interact. Instead, we shall assume that from time to time they collide, like a pair of tiny billiard balls, after which they go off at other speeds and in other directions that are subject to the laws of conservation of energy and momentum but otherwise random. Of course, it is not easy to see how they will do this if they are single points occupying no volume, but this part of the argument is needed only as an informal justification for some sort of randomness in the speeds and directions of the molecules. Maxwell's two very plausible assumptions about the nature of this randomness were that it should not change over time and that it should not distinguish between one direction and another. Roughly speaking, the second of these assumptions means that if d_1 and d_2 are two directions and s is a certain speed, then the chances that a particle is travelling at speed s in direction d_1 are the same as the chances that it is travelling at speed s in direction d_2. Surprisingly, these two assumptions are enough to determine exactly how the velocities should be distributed. That is, they tell us that if we want to choose the velocities randomly, then there is only one natural way to do it. (They should be assigned according to the normal distribution. This is the distribution that produces the famous 'bell curve', which occurs in a large number of different contexts, both mathematical and experimental.)

Once we have chosen the velocities, we can again forget all about interactions between the molecules. As a result, this slightly improved model shares many of the defects of the first one. In order to remedy them, there is no choice but to model the interactions somehow. It turns out that even very simple models of systems of interacting particles behave in a fascinating way and give rise to extremely difficult, indeed mostly unsolved, mathematical problems.

Modelling brains and computers

A computer can also be thought of as a collection of many simple parts that interact with one another, and largely for this reason theoretical computer science is also full of important unsolved problems. A good example of the sort of question one might like to answer is the following. Suppose that somebody chooses two prime numbers p and q, multiplies them together and tells you the answer pq. You can then work out what p and q are by taking every prime number in turn and seeing whether it goes exactly into pq. For example, if you are presented with the number 91, you can quickly establish that it is not a multiple of 2, 3, or 5, and then that it equals 7×13.

If, however, p and q are very large – with 200 digits each, say – then this process of trial and error takes an unimaginably long time, even with the help of a powerful computer. (If you want to get a feel for the difficulty, try finding two prime numbers that multiply to give 6901 and another two that give 280123.) On the other hand, it is not inconceivable that there is a much cleverer way to approach the problem, one that might be used as the basis for a computer program that does not take too long to run. If such a method could be found, it would allow one to break the codes on which most modern security systems are based, on the Internet and elsewhere, since the difficulty of deciphering these codes depends on the difficulty of factorizing large numbers. It would therefore be reassuring if there were some way of showing that a quick, efficient procedure for calculating p and q from their product pq does not exist. Unfortunately, while computers continually surprise us with what they can be used for, almost nothing is known about what they cannot do.

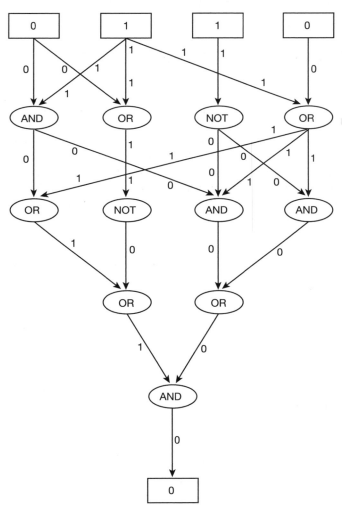

4. A primitive computer program

Before one can begin to think about this problem one must find some way of representing a computer mathematically, and as simply as possible. Figure 4 shows one of the best ways of doing this. It consists of layers of nodes that are linked to one another by lines that are called edges. Into the top layer goes the 'input', which is a sequence of 0s and 1s, and out of the bottom layer comes the 'output', which is another sequence of 0s and 1s. The nodes are of three kinds, called AND, OR, and NOT gates. Each of these gates receives some 0s and 1s from the edges that enter it from above. Depending on what it receives, it then sends out 0s or 1s itself, according to the following simple rules: if an AND gate receives nothing but 1s then it sends out 1s, and otherwise it sends out 0s; if an OR gate receives nothing but 0s then it sends out 0s, and otherwise it sends out 1s; only one edge is allowed to enter a NOT gate from above, and it sends out 1s if it receives a 0 and 0s if it receives a 1.

An array of gates linked by edges is called a *circuit*, and what I have described is the circuit model of computation. The reason 'computation' is an appropriate word is that a circuit can be thought of as taking one sequence of 0s and 1s and transforming it into another, according to some predetermined rules which may, if the circuit is large, be very complicated. This is also what computers do, although they translate these sequences out of and into formats that we can understand, such as high-level programming languages, windows, icons, and so on. There turns out to be a fairly simple way (from a theoretical point of view – it would be a nightmare to do in practice) of converting any computer program into a circuit that transforms 01-sequences according to exactly the same rules. Moreover, important characteristics of computer programs have their counterparts in the resulting circuits.

In particular, the number of nodes in the circuit corresponds to the length of time the computer program takes to run. Therefore, if one can show that a certain way of transforming 01-sequences needs a very large circuit, then one has also shown that it needs a computer program that runs for a very long time. The advantage of using the circuit model over analysing computers directly is that, from the mathematical point of view, circuits are simpler, more natural, and easier to think about.

A small modification to the circuit model leads to a useful model of the brain. Now, instead of 0s and 1s, one has signals of varying strengths that can be represented as numbers between 0 and 1. The gates, which correspond to neurons, or brain cells, are also different, but they still behave in a very simple way. Each one receives some signals from other gates. If the total strength of these signals – that is, the sum of all the corresponding numbers – is large enough, then the gate sends out its own signals of certain strengths. Otherwise, it does not. This corresponds to the decision of a neuron whether or not to 'fire'.

It may seem hard to believe that this model could capture the full complexity of the brain. However, that is partly because I have said nothing about how many gates there should be or how they should be arranged. A typical human brain contains about 100 billion neurons arranged in a very complicated way, and in the present state of knowledge about the brain it is not possible to say all that much more, at least about the fine detail. Nevertheless, the model provides a useful theoretical framework for thinking about how the brain might work, and it has allowed people to simulate certain sorts of brain-like behaviour.

Colouring maps and drawing up timetables

Suppose that you are designing a map that is divided into regions, and you wish to choose colours for the regions. You would like to use as few colours as possible, but do not wish to give two adjacent regions the same colour. Now suppose that you are drawing up the timetable for a university course that is divided into modules. The number of possible times for lectures is limited, so some modules will have to clash with others. You have a list of which students are taking which modules, and would like to choose the times in such a way that two modules clash only when there is nobody taking both.

These two problems appear to be quite different, but an appropriate choice of model shows that from the mathematical point of view they are the same. In both cases there are some objects (countries, modules) to which something must be assigned (colours, times). Some pairs of objects are incompatible (neighbouring countries, modules that must not clash) in the sense that they are not allowed to receive the same assignment. In neither problem do we really care what the objects are or what is being assigned to them, so we may as well just represent them as points. To show which pairs of points are incompatible we can link them with lines. A collection of points, some of which are joined by lines, is a mathematical structure known as a graph. Figure 5 gives a simple example. It is customary to call the points in a graph vertices, and the lines edges.

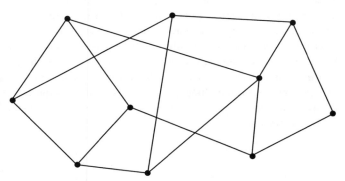

5. **A graph with 10 vertices and 15 edges**

Once we have represented the problems in this way, our task in both cases is to divide the vertices into a small number of groups in such a way that no group contains two vertices linked by an edge. (The graph in Figure 5 can be divided into three such groups, but not into two.) This illustrates another very good reason for making models as simple as possible: if you are lucky, the same model can be used to study many different phenomena at once.

Various meanings of the word 'abstract'

When devising a model, one tries to ignore as much as possible about the phenomenon under consideration, abstracting from it only those features that are essential to understanding its behaviour. In the examples I have discussed, stones were reduced to single points, the entire population of a country to one number, the brain to a network of gates obeying very simple mathematical rules, and the interactions between molecules to nothing at all. The resulting mathematical structures were abstract representations of the concrete situations being modelled.

These are two senses in which mathematics is an abstract subject: it abstracts the important features from a problem and it deals with objects that are not concrete and tangible. The next chapter will discuss a third, deeper sense of abstraction in mathematics, of which the example of the previous section has already given us some idea. A graph is a very flexible model with many uses. However, when one studies graphs, there is no need to bear these uses in mind: it does not matter whether the points represent regions, lectures, or something quite different again. A graph theorist can leave behind the real world entirely and enter the realm of pure abstraction.

Chapter 2
Numbers and abstraction

The abstract method

A few years ago, a review in the *Times Literary Supplement* opened with the following paragraph:

> Given that $0 \times 0 = 0$ and $1 \times 1 = 1$, it follows that there are numbers that are their own squares. But then it follows in turn that there are numbers. In a single step of artless simplicity, we seem to have advanced from a piece of elementary arithmetic to a startling and highly controversial philosophical conclusion: that numbers exist. You would have thought that it should have been more difficult.
>
> A. W. Moore reviewing *Realistic Rationalism*,
> by Jerrold J. Katz, in the *T.L.S.*, 11th September 1998.

This argument can be criticized in many ways, and it is unlikely that anybody takes it seriously, including the reviewer. However, there certainly are philosophers who take seriously the question of whether numbers exist, and this distinguishes them from mathematicians, who either find it obvious that numbers exist or do not understand what is being asked. The main purpose of this chapter is to explain why it is that mathematicians can, and even should, happily ignore this seemingly fundamental question.

The absurdity of the 'artlessly simple' argument for the existence of numbers becomes very clear if one looks at a parallel argument about the game of chess. Given that the black king, in chess, is sometimes allowed to move diagonally by one square, it follows that there are chess pieces that are sometimes allowed to move diagonally by one square. But then it follows in turn that there are chess pieces. Of course, I do not mean by this the mundane statement that people sometimes build chess sets – after all, it is possible to play the game without them – but the far more 'startling' philosophical conclusion that chess pieces exist independently of their physical manifestations.

What is the black king in chess? This is a strange question, and the most satisfactory way to deal with it seems to be to sidestep it slightly. What more can one do than point to a chessboard and explain the rules of the game, perhaps paying particular attention to the black king as one does so? What matters about the black king is not its existence, or its intrinsic nature, but the role that it plays in the game.

The abstract method in mathematics, as it is sometimes called, is what results when one takes a similar attitude to mathematical objects. This attitude can be encapsulated in the following slogan: a mathematical object *is* what it *does*. Similar slogans have appeared many times in the philosophy of language, and can be highly controversial. Two examples are 'In language there are only differences' and 'The meaning of a word is its use in the language', due to Saussure and Wittgenstein respectively (see Further reading), and one could add the rallying cry of the logical positivists: 'The meaning of a statement is its method of verification.' If you find mine unpalatable for philosophical reasons, then, rather than regarding it as a dogmatic pronouncement, think of it as an attitude which one can sometimes choose to adopt. In fact, as I hope to demonstrate, it is essential to be able to adopt it if one wants a proper understanding of higher mathematics.

Chess without the pieces

It is amusing to see, though my argument does not depend on it, that chess, or any other similar game, can be modelled by a graph. (Graphs were defined at the end of the previous chapter.) The vertices of the graph represent possible positions in the game. Two vertices P and Q are linked by an edge if the person whose turn it is to play in position P has a legal move that results in position Q. Since it may not be possible to get back from Q to P again, the edges need arrows on them to indicate their direction. Certain vertices are considered wins for white, and others wins for black. The game begins at one particular vertex, corresponding to the starting position of the game. Then the players take turns to move forwards along edges. The first player is trying to reach one of white's winning vertices, and the second one of black's. A far simpler game of this kind is illustrated in Figure 6. (It is not hard to see that for this game white has a winning strategy.)

This graph model of chess, though wildly impractical because of the vast number of possible chess positions, is perfect, in the sense that the resulting game is exactly equivalent to chess. And yet, when I defined it I made no mention of chess pieces at all. From this perspective, it seems quite extraordinary to ask whether the black

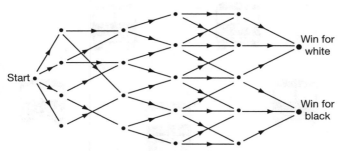

6. White starts, and has a winning strategy

king exists: the chessboard and pieces are nothing more than convenient organizing principles that help us think about the bewildering array of vertices and edges in a huge graph. If we say something like, 'The black king is in check', then this is just an abbreviation of a sentence that specifies an extremely long list of vertices and tells us that the players have reached one of them.

The natural numbers

'Natural' is the name given by mathematicians to the familiar numbers 1,2,3,4, ... They are among the most basic of mathematical objects, but they do not seem to encourage us to think abstractly. What, after all, could a number like 5 be said to *do*? It doesn't move around like a chess piece. Instead, it seems to have an intrinsic nature, a sort of pure fiveness that we immediately grasp when we look at a picture such as Figure 7.

However, when we consider larger numbers, there is rather less of this purity. Figure 8 gives us representations of the numbers 7, 12, and 47. Perhaps some people instantly grasp the sevenness of the first picture, but in most people's minds there will be a fleeting

7. The concept of fiveness

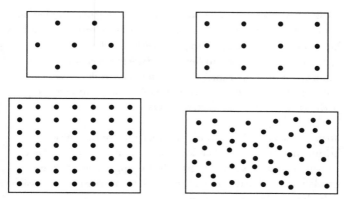

8. Ways of representing 7, 12, and 47 (twice)

thought such as, 'The outer dots form a hexagon, so together with the central one we get $6 + 1 = 7$.' Likewise, 12 will probably be thought of as 3×4, or 2×6. As for 47, there is nothing particularly distinctive about a group of that number of objects, as opposed to, say, 46. If they are arranged in a pattern, such as a 7×7 grid with two points missing, then we can use our knowledge that $7 \times 7 - 2 = 49 - 2 = 47$ to tell quickly how many there are. If not, then we have little choice but to count them, this time thinking of 47 as the number that comes after 46, which itself is the number that comes after 45, and so on.

In other words, numbers do not have to be very large before we stop thinking of them as isolated objects and start to understand them through their properties, through how they relate to other numbers, through their role in a *number system*. This is what I mean by what a number 'does'.

As is already becoming clear, the concept of a number is intimately connected with the arithmetical operations of addition and multiplication: for example, without some idea of arithmetic one

21

could have only the vaguest grasp of the meaning of a number like 1,000,000,017. A number system is not just a collection of numbers but a collection of numbers together with rules for how to do arithmetic. Another way to summarize the abstract approach is: think about the rules rather than the numbers themselves. The numbers, from this point of view, are tokens in a sort of game (or perhaps one should call them counters).

To get some idea of what the rules are, let us consider a simple arithmetical question: what does one do if one wants to become convinced that $38 \times 263 = 9994$? Most people would probably check it on a calculator, but if this was for some reason not possible, then they might reason as follows.

$$
\begin{aligned}
38 \times 263 &= 30 \times 263 + 8 \times 263 \\
&= 30 \times 200 + 30 \times 60 + 30 \times 3 + 8 \times 200 + 8 \times 60 + 8 \times 3 \\
&= 6000 + 1800 + 90 + 1600 + 480 + 24 \\
&= 9400 + 570 + 24 \\
&= 9994
\end{aligned}
$$

Why, though, do these steps seem so obviously correct? For example, why does one instantly believe that $30 \times 200 = 6000$? The definition of 30 is 3×10 and the definition of 200 is $2 \times (10 \times 10)$, so we can say with total confidence that $30 \times 200 = (3 \times 10) \times (2 \times (10 \times 10))$. But why is this 6000?

Normally, nobody would bother to ask this question, but to someone who did, we might say,

$$(3 \times 10) \times (2 \times (10 \times 10)) = (3 \times 2) \times (10 \times 10 \times 10) = 6 \times 1000 = 6000$$

Without really thinking about it, we would be using two familiar facts about multiplying: that if you multiply two numbers together, it doesn't matter which order you put them in, and that if you multiply more than two numbers together, then it makes no difference how you bracket them. For example, $7 \times 8 = 8 \times 7$ and

$(31 \times 34) \times 35 = 31 \times (34 \times 35)$. Notice that the intermediate calculations involved in the second of these two examples are definitely affected by the bracketing – but one knows that the final answer will be the same.

These two rules are called the *commutative* and *associative* laws for multiplication. Let me now list a few rules, including these two, that we commonly use when adding and multiplying.

A1 The commutative law for addition: $a + b = b + a$ for any two numbers a and b.

A2 The associative law for addition: $a + (b + c) = (a + b) + c$ for any three numbers a, b, and c.

M1 The commutative law for multiplication: $ab = ba$ for any two numbers a and b.

M2 The associative law for multiplication: $a(bc) = (ab)c$ for any three numbers a, b, and c.

M3 1 is a multiplicative identity: $1a = a$ for any number a.

D The distributive law: $(a + b)c = ac + bc$ for any three numbers a, b, and c.

I list these rules not because I want to persuade you that they are interesting on their own, but to draw attention to the role they play in our thinking, even about quite simple mathematical statements. Our confidence that $2 \times 3 = 6$ is probably based on a picture such as this.

$$* \qquad * \qquad *$$

$$* \qquad * \qquad *$$

On the other hand, a direct approach is out of the question if we want to show that $38 \times 263 = 9994$, so we think about this more complicated fact in an entirely different way, using the commutative, associative, and distributive laws. If we have obeyed these rules, then we believe the result. What is more, we believe it even if we have absolutely no visual sense of what 9994 objects would look like.

Zero

Historically, the idea of the number zero developed later than that of the positive integers. It has seemed to many people to be a mysterious and paradoxical concept, inspiring questions such as, 'How can something exist and yet be nothing?' From the abstract point of view, however, zero is very straightforward – it is just a new token introduced into our number system with the following special property.

A3 0 is an additive identity: $0 + a = a$ for any number a.

That is all you need to know about 0. Not what it means – just a little rule that tells you what it does.

What about other properties of the number 0, such as the fact that 0 times any number is 0? I did not list this rule, because it turns out that it can be deduced from property **A3** and our earlier rules. Here, for example, is how to show that $0 \times 2 = 0$, where 2 is defined to be the number $1 + 1$. First, rule **M1** tells us that $0 \times 2 = 2 \times 0$. Next, rule **D** tells us that $(1 + 1) \times 0 = 1 \times 0 + 1 \times 0$. But $1 \times 0 = 0$ by rule **M3**, so this equals $0 + 0$. Rule **A3** implies that $0 + 0 = 0$, and the argument is finished.

An alternative, non-abstract argument might be something like this: '0×2 means *add up no twos*, and if you do that you are left with nothing, that is, 0.' But this way of thinking makes it hard to answer questions such as the one asked by my son John (when six): how can nought times nought be nought, since nought times nought means that you have *no* noughts? A good answer, though not one that was suitable at the time, is that it can be deduced from the rules as follows. (After each step, I list the rule I am using.)

$$
\begin{aligned}
0 &= 1 \times 0 && \textbf{M3} \\
  &= (0 + 1) \times 0 && \textbf{A3} \\
  &= 0 \times 0 + 1 \times 0 && \textbf{D} \\
  &= 0 \times 0 + 0 && \textbf{M3} \\
  &= 0 + 0 \times 0 && \textbf{A1} \\
  &= 0 \times 0 && \textbf{A3}
\end{aligned}
$$

Why am I giving these long-winded proofs of very elementary facts? Again, it is not because I find the proofs mathematically interesting, but rather because I wish to show what it means to justify arithmetical statements abstractly (by using a few simple rules and not worrying what numbers actually are) rather than concretely (by reflecting on what the statements mean). It is of course very useful to associate meanings and mental pictures with mathematical objects, but, as we shall see many times in this book, often these associations are not enough to tell us what to do in new and unfamiliar contexts. Then the abstract method becomes indispensable.

Negative numbers and fractions

As anybody with experience of teaching mathematics to small children knows, there is something indirect about subtraction and division that makes them harder to understand than addition and multiplication. To explain subtraction, one can of course use the notion of taking away, asking questions such as, 'How many oranges will be left if you start with five and eat two of them?'. However, that is not always the best way to think about it. For example, if one has to subtract 98 from 100, then it is better to think not about taking 98 away from 100, but about what one has to add to 98 to make 100. Then, what one is effectively doing is solving the equation $98 + x = 100$, although of course it is unusual for the letter x actually to cross one's mind during the calculation. Similarly, there are two ways of thinking about division. To explain the meaning of 50 divided by 10, one can either ask, 'If fifty objects are split into ten equal groups, then how many will be in each group?' or ask, 'If fifty objects are split into groups of ten, then how many groups will there be?'. The second approach is equivalent to the question, 'What must ten be multiplied by to make fifty, which in turn is equivalent to solving the equation $10x = 50$.

A further difficulty with explaining subtraction and division to children is that they are not always possible. For example, you cannot take ten oranges away from a bowl of seven, and three children cannot share eleven marbles equally. However, that does not stop adults subtracting 10 from 7 or dividing 11 by 3, obtaining the answers -3 and $11/3$ respectively. The question then arises: do the numbers -3 and $11/3$ actually exist, and if so what are they?

From the abstract point of view, we can deal with these questions as we dealt with similar questions about zero: by forgetting about them. All we need to know about -3 is that when you add 3 to it you get 0, and all we need to know about $11/3$ is that when you multiply it by 3 you get 11. Those are the rules, and, in conjunction with earlier rules, they allow us to do arithmetic in a larger number system. Why should we wish to extend our number system in this way? Because it gives us a model in which equations like $x + a = b$ and $ax = b$ can be solved, whatever the values of a and b, except that a should not be 0 in the second equation. To put this another way, it gives us a model where subtraction and division are always possible, as long as one does not try to divide by 0. (The issue of division by 0 will be discussed later in the chapter.)

As it happens, we need only two more rules to extend our number system in this way: one that gives us negative numbers and one that gives us fractions, or *rational* numbers as they are customarily known.

A4 Additive inverses: for every number a there is a number b such that $a + b = 0$.

M4 Multiplicative inverses: for every number a apart from 0 there is a number c such that $ac = 1$.

Armed with these rules, we can think of $-a$ and $1/a$ as notation for the numbers b in **A4** and c in **M4** respectively. As for a more general expression like p/q, it stands for p multiplied by $1/q$.

The rules **A4** and **M4** imply two further rules, known as cancellation laws.

A5 The cancellation law for addition: if a, b, and c are any three numbers and $a + b = a + c$, then $b = c$.
M5 The cancellation law for multiplication: if a, b, and c are any three numbers, a is not 0 and $ab = ac$, then $b = c$.

The first of these is proved by adding $-a$ to both sides, and the second by multiplying both sides by $1/a$, just as you would expect. Note the different status of **A5** and **M5** from the rules that have gone before – they are *consequences* of the earlier rules, rather than rules we simply introduce to make a good game.

If one is asked to add two fractions, such as $2/5$ and $3/7$, then the usual method is to give them a common denominator, as follows:

$$\frac{2}{5} + \frac{3}{7} = \frac{14}{35} + \frac{15}{35} = \frac{29}{35}$$

This technique, and others like it, can be justified using our new rules. For example,

$$35 \times \frac{14}{35} = 35 \times \left(14 \times \frac{1}{35}\right) = (35 \times 14) \times \frac{1}{35} = (14 \times 35) \times \frac{1}{35}$$
$$= 14 \times \left(35 \times \frac{1}{35}\right) = 14 \times 1 = 14,$$

and

$$35 \times \frac{2}{5} = (5 \times 7) \times \left(2 \times \frac{1}{5}\right) = (7 \times 5) \times \left(\frac{1}{5} \times 2\right) = \left(7 \times \left(5 \times \frac{1}{5}\right)\right) \times 2$$
$$= (7 \times 1) \times 2 = 7 \times 2 = 14.$$

Hence, by rule **M5**, 2/5 and 14/35 are equal, as we assumed in the calculation.

Similarly, one can justify familiar facts about negative numbers. I leave it to the reader to deduce from the rules that $(-1) \times (-1) = 1$ – the deduction is fairly similar to the proof that $0 \times 0 = 0$.

Why does it seem to many people that negative numbers are less real than positive ones? Probably it is because counting small groups of objects is a fundamental human activity, and when we do it we do not use negative numbers. But all this means is that the natural number system, thought of as a model, is useful in certain circumstances where an enlarged number system is not. If we want to think about temperatures, or dates, or bank accounts, then negative numbers *do* become useful. As long as the extended number system is logically consistent, which it is, there is no harm in using it as a model.

It may seem strange to call the natural number system a model. Don't we *actually count*, with no particular idealization involved? Yes we do, but this procedure is not always appropriate, or even possible. There is nothing wrong with the number 139484027593649864923987 from a mathematical point of view, but if we can't even count votes in Florida, it is inconceivable that we might ever be sure that we had a collection of 139484027593649864923987 objects. If you take two piles of leaves and add them to a third, the result is not three piles of leaves but one large pile. And if you have just watched a rainstorm, then, as Wittgenstein said, 'The proper answer to the question, "How many drops did you see?", is *many*, not that there was a number but you don't know how many.'

Real and complex numbers

The real number system consists of all numbers that can be represented by infinite decimals. This concept is more sophisticated than it seems, for reasons that will be explained in Chapter 4. For now, let me say that the reason for extending our number system from rational to real numbers is similar to the reason for introducing negative numbers and fractions: they allow us to solve equations that we could not otherwise solve.

The most famous example of such an equation is $x^2 = 2$. In the sixth century BC it was discovered by the school of Pythagoras that $\sqrt{2}$ is irrational, which means that it cannot be represented by a fraction. (A proof of this will be given in the next chapter.) This discovery caused dismay when it was made, but now we cheerfully accept that we must extend our number system if we want to model things like the length of the diagonal of a square. Once again, the abstract method makes our task very easy. We introduce a new symbol, $\sqrt{2}$, and have one rule that tells us what to do with it: it squares to 2.

If you are well trained, then you will object to what I have just said on the grounds that the rule does not distinguish between $\sqrt{2}$ and $-\sqrt{2}$. One way of dealing with this is to introduce a new concept into our number system, that of *order*. It is often useful to talk of one number being bigger than another, and if we allow ourselves to do that then we can pick out $\sqrt{2}$ by the additional property that $\sqrt{2}$ is greater than 0. But even without this property we can do calculations such as

$$\frac{1}{\sqrt{2}-1} = \frac{\sqrt{2}+1}{(\sqrt{2}-1)(\sqrt{2}+1)} = \frac{\sqrt{2}+1}{(\sqrt{2})^2 - \sqrt{2} + \sqrt{2} - 1} = \frac{\sqrt{2}+1}{2-1} = \sqrt{2}+1,$$

and there is actually an advantage in *not* distinguishing between $\sqrt{2}$ and $-\sqrt{2}$, which is that then we know that the above calculation is valid for both numbers.

Historical suspicion of the abstract method has left its traces in the words used to describe the new numbers that arose each time the number system was enlarged, words like 'negative' and 'irrational'. But far harder to swallow than these were the 'imaginary', or 'complex', numbers, that is, numbers of the form $a + bi$, where a and b are real numbers and i is the square root of -1.

From a concrete point of view, one can quickly dismiss the square root of -1: since the square of any number is positive, -1 does not have a square root, and that is the end of the story. However, this objection carries no force if one adopts an abstract attitude. Why not simply continue to extend our number system, by introducing a solution to the equation $x^2 = -1$ and calling it i? Why should this be more objectionable than our earlier introduction of $\sqrt{2}$?

One answer might be that $\sqrt{2}$ has a decimal expansion which can (in principle) be calculated to any desired accuracy, while nothing equivalent can be said about i. But all this says is something we already know, namely that i is not a real number – just as $\sqrt{2}$ is not a rational number. It does not stop us extending our number system to one in which we can do calculations such as

$$\frac{1}{i-1} = \frac{i+1}{(i-1)(i+1)} = \frac{i+1}{i^2 - i + i - 1} = \frac{i+1}{-1-1} = -\frac{1}{2}(i+1)$$

The main difference between i and $\sqrt{2}$ is that in the case of i we are *forced* to think abstractly, whereas there is always the option with $\sqrt{2}$ of using a concrete representation such as 1.4142 . . . or the length of the diagonal of a unit square. To see why i has no such representation, ask yourself the following question: which of the two square roots of -1 is i and which is $-i$? The question does not make sense because the *only* defining property of i is that it squares to -1. Since $-i$ has the same property, any true sentence about i remains true if one replaces it with the corresponding sentence about $-i$. It is difficult, once one has grasped this, to have any respect for the view that i might denote an independently existing Platonic object.

There is a parallel here with a well-known philosophical conundrum. Might it be that when you perceive the colour red your sensation is what I experience when I perceive green, and vice versa? Some philosophers take this question seriously and define 'qualia' to be the absolute intrinsic experiences we have when, for example, we see colours. Others do not believe in qualia. For them, a word like 'green' is defined more abstractly by its role in a linguistic system, that is, by its relationships with concepts like 'grass', 'red', and so on. It is impossible to deduce somebody's position on this issue from the way they talk about colour, except during philosophical debates. Similarly, all that matters in practice about numbers and other mathematical objects is the rules they obey.

If we introduced i in order to have a solution to the equation $x^2 = -1$, then what about other, similar equations such as $x^4 = -3$, or $2x^6 + 3x + 17 = 0$? Remarkably, it turns out that every such equation can be solved within the complex number system. In other words, we make the small investment of accepting the number i, and are then repaid many times over. This fact, which has a complicated history but is usually attributed to Gauss, is known as the fundamental theorem of algebra and it provides very convincing evidence that there is something natural about i. It may be impossible to imagine a basket of i apples, a car journey that lasts i hours, or a bank account with an overdraft of i pounds, but the complex number system has become indispensable to mathematicians, and to scientists and engineers as well – the theory of quantum mechanics, for example, depends heavily on complex numbers. It provides one of the best illustrations of a general principle: if an abstract mathematical construction is sufficiently natural, then it will almost certainly find a use as a model.

A first look at infinity

Once one has learned to think abstractly, it can be exhilarating, a bit like suddenly being able to ride a bicycle without having to worry about keeping one's balance. However, I do not wish to give the impression that the abstract method is like a licence to print money. It is interesting to contrast the introduction of i into the number system with what happens when one tries to introduce the number infinity. At first there seems to be nothing stopping us: infinity should mean something like 1 divided by 0, so why not let ∞ be an abstract symbol and regard it as a solution to the equation $0x = 1$?

The trouble with this idea emerges as soon as one tries to do arithmetic. Here, for example, is a simple consequence of **M2**, the associative law for multiplication, and the fact that $0 \times 2 = 0$.

$$1 = \infty \times 0 = \infty \times (0 \times 2) = (\infty \times 0) \times 2 = 1 \times 2 = 2$$

What this shows is that the existence of a solution to the equation $0x = 1$ leads to an *inconsistency*. Does that mean that infinity does not exist? No, it simply means that no natural notion of infinity is compatible with the laws of arithmetic. It is sometimes useful to extend the number system to include the symbol ∞, accepting that in the enlarged system these laws are not always valid. Usually, however, one prefers to keep the laws and do without infinity.

Raising numbers to negative and fractional powers

One of the greatest virtues of the abstract method is that it allows us to make sense of familiar concepts in unfamiliar situations. The phrase 'make sense' is quite appropriate, since that is just what we do, rather than discovering some pre-existing sense. A simple example of this is the way we extend the concept of raising a number to a power.

If n is a positive integer, then a^n means the result of multiplying n as together. For example, $5^3 = 5 \times 5 \times 5 = 125$ and $2^5 = 2 \times 2 \times 2 \times 2 \times 2 = 32$. But with this as a definition, it is not easy to interpret an expression such as $2^{3/2}$, since you cannot take one and a half twos and multiply them together. What is the abstract method for dealing with a problem like this? Once again, it is not to look for intrinsic meaning – in this case of expressions like a^n – but to think about rules.

Two elementary rules about raising numbers to powers are the following.

E1 $a^1 = a$ for any real number a.
E2 $a^{m+n} = a^m \times a^n$ for any real number a and any pair of natural numbers m and n.

For example, $2^5 = 2^3 \times 2^2$ since 2^5 means $2 \times 2 \times 2 \times 2 \times 2$ and $2^3 \times 2^2$ means $(2 \times 2 \times 2) \times (2 \times 2)$. These are the same number because multiplication is associative.

From these two rules we can quickly recover the facts we already know. For example, $a^2 = a^{1+1}$ which, by **E2**, is $a^1 \times a^1$. By **E1** this is $a \times a$, as it should be. But we are now in a position to do much more. Let us write x for the number $2^{3/2}$. Then $x \times x = 2^{3/2} \times 2^{3/2}$ which, by **E2**, is $2^{3/2 + 3/2} = 2^3 = 8$. In other words, $x^2 = 8$. This does not quite determine x, since 8 has two square roots, so it is customary to adopt the following convention.

E3 If $a > 0$ and b is a real number, then a^b is positive.

Using **E3** as well, we find that $2^{3/2}$ is the positive square root of 8.

This is not a discovery of the 'true value' of $2^{3/2}$. However, neither is the interpretation we have given to the expression $2^{3/2}$ arbitrary – it is the only possibility if we want to preserve rules **E1**, **E2**, and **E3**.

A similar argument allows us to interpret a^0, at least when a is not 0. By **E1** and **E2** we know that $a = a^1 = a^{1+0} = a^1 \times a^0 = a \times a^0$. The cancellation law **M5** then implies that $a^0 = 1$, whatever the value of a. As for negative powers, if we know the value of a^b, then $1 = a^0 = a^{b+(-b)} = a^b \times a^{-b}$, from which it follows that $a^{-b} = 1/a^b$. The number $2^{-3/2}$, for example, is $1/\sqrt{8}$.

Another concept that becomes much simpler when viewed abstractly is that of a logarithm. I will not have much to say about logarithms in this book, but if they worry you, then you may be reassured to learn that all you need to know in order to use them are the following three rules. (If you want logarithms to base e instead of 10, then just replace 10 by e in **L1**.)

L1 $\log(10) = 1$.
L2 $\log(xy) = \log(x) + \log(y)$.
L3 If $x < y$ then $\log(x) < \log(y)$.

For example, to see that $\log(30)$ is less than $3/2$, note that

$$\log(1000) = \log(10) + \log(100) = \log(10) + \log(10) + \log(10) = 3,$$

by **L1** and **L2**. But $2\log(30) = \log(30) + \log(30) = \log(900)$, by **L2**, and $\log(900) < \log(1000)$, by **L3**. Hence, $2\log(30) < 3$, so that $\log(30) < 3/2$.

I shall discuss many concepts, later in the book, of a similar nature to these. They are puzzling if you try to understand them concretely, but they lose their mystery when you relax, stop worrying about what they *are*, and use the abstract method.

Chapter 3
Proofs

The diagram below shows five circles, the first with one point on its boundary, the second with two, and so on. All the possible lines joining these boundary points have also been drawn, and these lines divide the circles into regions. If one counts the regions in each circle one obtains the sequence 1,2,4,8,16. This sequence is instantly recognizable: it seems that the number of regions doubles each time

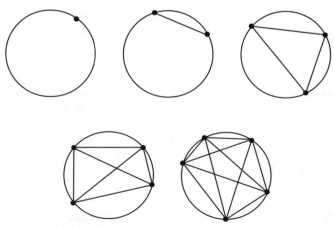

9. **Dividing a circle into regions**

a new point is added to the boundary, so that n points define 2^{n-1} regions, at least if no three lines meet at a point.

Mathematicians, however, are rarely satisfied with the phrase 'it seems that'. Instead, they demand a *proof*, that is, an argument that puts a statement beyond all possible doubt. What, though, does this mean? Although one can often establish a statement as true beyond all *reasonable* doubt, surely it is going a bit far to claim that an argument leaves no room for doubt whatsoever. Historians can provide many examples of statements, some of them mathematical, that were once thought to be beyond doubt, but which have since been shown to be incorrect. Why should the theorems of present-day mathematics be any different? I shall answer this question by giving a few examples of proofs and drawing from them some general conclusions.

The irrationality of the square root of two

As I mentioned in the last chapter, a number is called rational if it can be written as a fraction p/q, where p and q are whole numbers, and irrational if it cannot. One of the most famous proofs in mathematics shows that $\sqrt{2}$ is irrational. It illustrates a technique known as *reductio ad absurdum*, or proof by contradiction.

A proof of this kind begins with the assumption that the result to be proved is *false*. This may seem a strange way to argue, but in fact we often use the same technique in everyday conversation. If you went to a police station to report that you had seen a car being vandalized and were accused of having been the vandal yourself, you might well say, 'If *I* had done it, I would hardly be drawing attention to myself in this way.' You would be temporarily entertaining the (untrue) hypothesis that you were the vandal, in order to show how ridiculous it was.

We are trying to prove that $\sqrt{2}$ is irrational, so let us assume that it is *rational* and try to show that this assumption leads to absurd consequences. I shall write the argument out as a sequence of steps, giving more detail than many readers are likely to need.

1. If $\sqrt{2}$ is rational, then we can find whole numbers p and q such that $\sqrt{2} = p/q$ (by the definition of 'rational').

2. Any fraction p/q is equal to some fraction r/s where r and s are not both even. (Just keep dividing the top and bottom of the fraction by 2 until at least one of them becomes odd. For example, the fraction 1412/1000 equals 706/500 equals 353/250.)

3. Therefore, if $\sqrt{2}$ is rational, then we can find whole numbers r and s, not both even, such that $\sqrt{2} = r/s$.

4. If $\sqrt{2} = r/s$, then $2 = r^2/s^2$ (squaring both sides of the equation).

5. If $2 = r^2/s^2$, then $2s^2 = r^2$ (multiplying both sides by s^2).

6. If $2s^2 = r^2$, then r^2 is even, which means that r must be even.

7. If r is even, then $r = 2t$ for some whole number t (by the definition of 'even').

8. If $2s^2 = r^2$ and $r = 2t$, then $2s^2 = (2t)^2 = 4t^2$, from which it follows that $s^2 = 2t^2$ (dividing both sides by 2).

9. If $s^2 = 2t^2$, then s^2 is even, which means that s is even.

10. Under the assumption that $\sqrt{2}$ is rational, we have shown that $\sqrt{2} = r/s$, with r and s not both even (step 3). We have then shown that r is even (step 6) and that s is even (step 9). This is a clear contradiction. Since the assumption that $\sqrt{2}$ is rational has consequences that are clearly false, the assumption itself must be false. Therefore, $\sqrt{2}$ is irrational.

I have tried to make each of the above steps so obviously valid that the conclusion of the argument is undeniable. However, have I really left *no* room for doubt? If somebody offered you the chance of ten thousand pounds, on condition that you would have to forfeit your life if two positive whole numbers p and q were ever discovered such that $p^2 = 2q^2$, then would you accept the offer? If so, would you be in the slightest bit worried?

Step 6 contains the assertion that if r^2 is even, then r must also be even. This seems pretty obvious (an odd number times an odd number is odd) but perhaps it could do with more justification if we are trying to establish *with absolute certainty* that $\sqrt{2}$ is irrational. Let us split it into five further substeps:

6a. r is a whole number and r^2 is even. We would like to show that r must also be even. Let us assume that r is odd and seek a contradiction.

6b. Since r is odd, there is a whole number t such that $r = 2t + 1$.

6c. It follows that $r^2 = (2t + 1)^2 = 4t^2 + 4t + 1$.

6d. But $4t^2 + 4t + 1 = 2(2t^2 + 2t) + 1$, which is odd, contradicting the fact that r^2 is even.

6e. Therefore, r is even.

Does this now make step 6 completely watertight? Perhaps not, because substep 6b needs to be justified. After all, the definition of an odd number is simply a whole number that is not a multiple of two. Why should every whole number be either a multiple of two or one more than a multiple of two? Here is an argument that establishes this.

6b1. Let us call a whole number r *good* if it is either a multiple of two or one more than a multiple of two. If r is good, then $r = 2s$ or $r = 2s + 1$, where s is also a whole number. If $r = 2s$ then $r + 1 = 2s + 1$, and if $r = 2s + 1$, then $r + 1 = 2s + 2 = 2(s + 1)$. Either way, $r + 1$ is also good.

6b2. 1 is good, since $0 = 0 \times 2$ is a multiple of 2 and $1 = 0 + 1$.

6b3. Applying step 6b1 repeatedly, we can deduce that 2 is good, then that 3 is good, then that 4 is good, and so on.

6b4. Therefore, every positive whole number is good, as we were trying to prove.

Have we now finished? Perhaps the shakiest step this time is

6b4, because of the rather vague words 'and so on' from the previous step. Step 6b3 shows us how to demonstrate, for any *given* positive whole number n, that it is good. The trouble is that in the course of the argument we will have to count from 1 to n, which, if n is large, will take a very long time. The situation is even worse if we are trying to show that *every* positive whole number is good. Then it seems that the argument will never end.

On the other hand, given that steps 6b1 to 6b3 genuinely and unambiguously provide us with a method for showing that any individual n is good (provided that we have time), this objection seems unreasonable. So unreasonable, in fact, that mathematicians adopt the following principle as an axiom.

> Suppose that for every positive integer n there is associated a statement $S(n)$. (In our example, $S(n)$ stands for the statement 'n is good'.) If $S(1)$ is true, and if the truth of $S(n)$ always implies the truth of $S(n + 1)$, then $S(n)$ is true for every n.

This is known as the principle of mathematical induction, or just induction to those who are used to it. Put less formally, it says that if you have an infinite list of statements that you wish to prove, then one way to do it is to show that the first one is true and that each one implies the next.

As the last few paragraphs illustrate, the steps of a mathematical argument can be broken down into smaller and therefore more clearly valid substeps. These steps can then be broken down into subsubsteps, and so on. A fact of fundamental importance to mathematics is that this process *eventually comes to an end*. In principle, if you go on and on splitting steps into smaller ones, you will end up with a very long argument starts with axioms that are universally accepted and proceeds to the desired conclusion by means of only the most elementary logical rules (such as 'if A is true and A implies B then B is true').

What I have just said in the last paragraph is far from obvious: in fact it was one of the great discoveries of the early 20th century, largely due to Frege, Russell, and Whitehead (see Further reading). This discovery has had a profound impact on mathematics, because it means that *any dispute about the validity of a mathematical proof can always be resolved*. In the 19th century, by contrast, there were genuine disagreements about matters of mathematical substance. For example, Georg Cantor, the father of modern set theory, invented arguments that relied on the idea that one infinite set can be 'bigger' than another. These arguments are accepted now, but caused great suspicion at the time. Today, if there is disagreement about whether a proof is correct, it is either because the proof has not been written in sufficient detail, or because not enough effort has been spent on understanding it and checking it carefully.

Actually, this does not mean that disagreements never occur. For example, it quite often happens that somebody produces a very long proof that is unclear in places and contains many small mistakes, but which is not obviously incorrect in a fundamental way. Establishing conclusively whether such an argument can be made watertight is usually extremely laborious, and there is not much reward for the labour. Even the author may prefer not to risk finding that the argument is wrong.

Nevertheless, the fact that disputes can *in principle* be resolved does make mathematics unique. There is no mathematical equivalent of astronomers who still believe in the steady-state theory of the universe, or of biologists who hold, with great conviction, very different views about how much is explained by natural selection, or of philosophers who disagree fundamentally about the relationship between consciousness and the physical world, or of economists who follow opposing schools of thought such as monetarism and neo-Keynesianism.

It is important to understand the phrase 'in principle' above. No mathematician would ever bother to write out a proof in complete detail – that is, as a deduction from basic axioms using only the most utterly obvious and easily checked steps. Even if this were feasible it would be quite unnecessary: mathematical papers are written for highly trained readers who do not need everything spelled out. However, if somebody makes an important claim and other mathematicians find it hard to follow the proof, they will ask for clarification, and the process will then begin of dividing steps of the proof into smaller, more easily understood substeps. Usually, again because the audience is highly trained, this process does not need to go on for long until either the necessary clarification has been provided or a mistake comes to light. Thus, a purported proof of a result that other mathematicians care about is almost always accepted as correct only if it *is* correct.

I have not dealt with a question that may have occurred to some readers: why should one accept the axioms proposed by mathematicians? If, for example, somebody were to object to the principle of mathematical induction, how could the objection be met? Most mathematicians would give something like the following response. First, the principle seems obviously valid to virtually everybody who understands it. Second, what matters about an axiom system is less the *truth* of the axioms than their consistency and their usefulness. What a mathematical proof actually does is show that certain conclusions, such as the irrationality of $\sqrt{2}$, follow from certain premises, such as the principle of mathematical induction. The validity of these premises is an entirely independent matter which can safely be left to philosophers.

The irrationality of the golden ratio

A common experience for people learning advanced mathematics is to come to the end of a proof and think, 'I understood how each line followed from the previous one, but somehow I am none the wiser about *why* the theorem is true, or how anybody thought of this

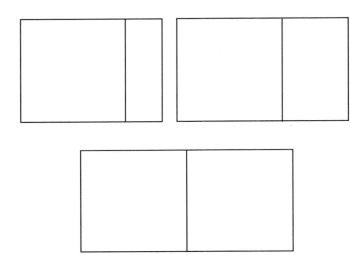

10. The existence of the golden ratio

argument'. We usually want more from a proof than a mere guarantee of correctness. We feel after reading a good proof that it provides an explanation of the theorem, that we understand something we did not understand before.

Since a large proportion of the human brain is devoted to the processing of visual data, it is not surprising that many arguments exploit our powers of visualization. To illustrate this, I shall give another proof of irrationality, this time of the so-called golden ratio. This is a number that has fascinated non-mathematicians (and to a lesser extent mathematicians) for centuries. It is the ratio of the side lengths of a rectangle with the following property: if you cut a square off it then you are left with a smaller, rotated rectangle of exactly the same shape as the original one. This is true of the second rectangle in Figure 10.

Why should such a ratio exist at all? (Mathematicians are trained to ask this sort of question.) One way to see it is to imagine a small rectangle growing out of the side of a square so that the square turns into a larger rectangle. To begin with, the small rectangle is very long and thin, while the larger one is still almost a square. If we allow the small rectangle to grow until it becomes a square itself, then the larger rectangle has become twice as long as it is wide. Thus, at first the smaller rectangle was much thinner than the larger one, and now it is fatter (relative to its size). Somewhere in between there must be a point where the two rectangles have the same shape. Figure 10 illustrates this process.

A second way of seeing that the golden ratio exists is to calculate it. If we call it x and assume that the square has side lengths 1, then the side lengths of the large rectangle are 1 and x, while the side lengths of the small one are $x - 1$ and 1. If they are the same shape, then $x = \dfrac{x}{1} = \dfrac{1}{x-1}$. Multiplying both sides by $x - 1$ we deduce that $x(x - 1) = 1$, so $x^2 - x - 1 = 0$. Solving this quadratic equation, and bearing in mind that x is not a negative number, we find that $x = \dfrac{1 + \sqrt{5}}{2}$. (If you are particularly well trained mathematically, or have taken the last chapter to heart, you may now ask why I am so confident that $\sqrt{5}$ exists. In fact, what this second argument does is to reduce a geometrical problem to an equivalent algebraic one.)

Having established that the ratio x exists, let us take a rectangle with sides of length x and 1 and consider the following process. First, cut off a square from it, leaving a smaller rectangle which, by the definition of the golden ratio, has the same shape as the original one. Now repeat this basic operation over and over again, obtaining a sequence of smaller and smaller rectangles, each of the same shape as the one before and hence each with side lengths in the golden ratio. Clearly, the process will never end. (See the first rectangle of Figure 11.)

Now let us do the same to a rectangle with side lengths in the ratio p/q, where p and q are whole numbers. This means that the

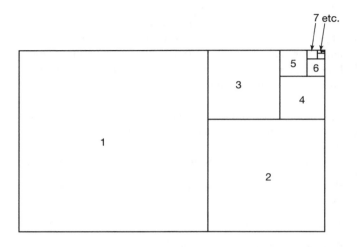

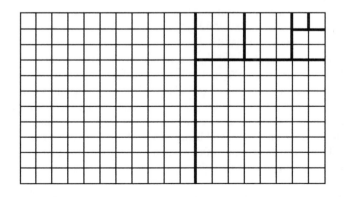

11. Squares are removed from the rectangle in the order shown

rectangle has the same shape as a rectangle with side lengths p and q, so it can be divided into $p \times q$ little squares, as illustrated by the second rectangle of Figure 11. What happens if we remove large squares from the end of this rectangle? If q is smaller than p, then we will remove a $q \times q$ square and end up with a $q \times (p - q)$ rectangle. We can then remove a further square, and so on. Can the process go on for ever? No, because each time we cut off a square we remove a whole number of the little squares, and we cannot possibly do this more than $p \times q$ times because there were only $p \times q$ little squares to start with.

We have shown the following two facts.

1. If the ratio of the sides of the rectangle is the golden ratio, then one can continue cutting off squares for ever.
2. If the ratio of the sides of the rectangle is p/q for some pair of whole numbers p and q, then one cannot continue cutting off squares for ever.

It follows that the ratio p/q is not the golden ratio, whatever the values of p and q. In other words, the golden ratio is irrational.

If you think very hard about the above proof, you will eventually realize that it is not as different from the proof of the irrationality of $\sqrt{2}$ as it might at first appear. Nevertheless, the way it is presented is certainly different – and for many people more appealing.

Regions of a circle

Now that I have said something about the nature of mathematical proof, let us return to the problem with which the chapter began. We have a circle with n points round its boundary, we join all pairs of these points with straight lines, and we wish to show that the number of regions bounded by these lines will be 2^{n-1}. We have already seen that this is true if n is 1, 2, 3, 4, or 5. In order to prove the statement in general, we would very much like to find a convincing *reason* for the number of regions to double each time a new point is added to the boundary. What could such a reason be?

Nothing immediately springs to mind, so one way of getting started might be to study the diagrams of the divided-up circles and see whether we notice some pattern that can be generalized. For example, three points round the boundary produce three outer regions and one central one. With four points, there are four outer regions and four inner ones. With five points, there is a central pentagon, with five triangles pointing out of it, five triangles slotting into the resulting star and making it back into a pentagon, and finally five outer regions. It therefore seems natural to think of 4 as 3 + 1, of 8 as 4 + 4, and of 16 as 5 + 5 + 5 + 1.

Does this help? We do not seem to have enough examples for a clear pattern to emerge, so let us try drawing the regions that result from six points placed round the boundary. The result appears in Figure 12. Now there are six outer regions. Each of these is next to a triangular region that points inwards. Between two neighbouring regions of this kind are two smaller triangular regions. So far we have 6 + 6 + 12 = 24 regions and have yet to count the regions inside the central hexagon. These split into three pentagons, three quadrilaterals, and one central triangle. It therefore seems natural to think of the number of regions as 6 + 6 + 12 + 3 + 3 + 1.

Something seems to be wrong, though, because this gives us 31. Have we made a mistake? As it happens, no: the correct sequence begins 1, 2, 4, 8, 16, 31, 57, 99, 163. In fact, with a little further reflection one can see that the number of regions *could not possibly* double every time. For a start, it is worrying that the number of regions defined when there are 0 points round the boundary is 1 rather than 1/2, which is what it would have to be if it doubled when the first point was put in. Though anomalies of this kind sometimes happen with zero, most mathematicians would find this particular one troubling. However, a more serious problem is that if *n* is a

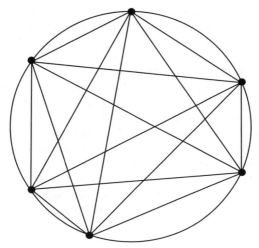

12. Regions of a circle

fairly large number, then 2^{n-1} is quite obviously too big. For example, 2^{n-1} is 524,288 when $n = 20$ and 536,870,912 when $n = 30$. Is it remotely plausible that 30 points round the edge of a circle would define over five hundred million different regions? Surely not. Imagine drawing a large circle in a field, pushing thirty pegs into it at irregularly spaced intervals, and then joining them with very thin string. The number of resulting regions would certainly be quite large, but not unimaginably so. If the circle had a diameter of ten metres and was divided into five hundred million regions, then there would have to be, on average, over six hundred regions per square centimetre. The circle would have to be thick with string, but with only thirty points round the boundary it clearly wouldn't be.

As I said earlier, mathematicians are wary of words like 'clearly'. However, in this instance our intuition can be backed up by a solid argument, which can be summarized as follows. If the circle is divided into a vast number of polygonal regions, then these regions must have, between them, a vast number of corners. Each corner is a point where two pieces of string cross, and to each such crossing one can associate four pegs, namely the ones where the relevant pieces of string end. There are 30 possible choices for the first peg, 29 for the second, 28 for the third, and 27 for the fourth. This suggests that the number of ways of choosing the four pegs is $30 \times 29 \times 28 \times 27 = 657720$, but that is to forget that if we had chosen the same four pegs in a different order, then we would have specified the same crossing. There are $4 \times 3 \times 2 \times 1 = 24$ ways of putting any given four pegs in order, and if we allow for this we find that the number of crossings is $657720/24 = 27405$, which is nothing like vast enough to be the number of corners of 536,870,912 regions. (In fact, the true number of regions produced by 30 points turns out to be 31,931.)

This cautionary tale contains many important lessons about the justification of mathematical statements. The most obvious one is that if you do not take care to prove what you say, then you run the risk of saying something that is wrong. A more positive moral is that if you do try to prove statements, then you will understand them in a completely different and much more interesting way.

Pythagoras' theorem

The famous theorem of Pythagoras states that if a right-angled triangle has sides of length a, b, and c, where c is the length of the hypotenuse (the side opposite the right angle), then $a^2 + b^2 = c^2$. It has several proofs, but one stands out as particularly short and easy to understand. Indeed, it needs little more than the following two diagrams.

In Figure 13, the squares that I have labelled A, B, and C have sides of length a, b, and c respectively, and therefore areas a^2, b^2, and c^2. Since moving the four triangles does not change their area or cause them to overlap, the area of the part of the big square that they do

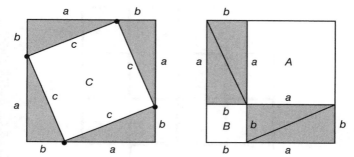

13. A short proof of Pythagoras' theorem

not cover is the same in both diagrams. But on the left this area is $a^2 + b^2$ and on the right it is c^2.

Tiling a square grid with the corners removed

Here is a well-known brainteaser. Take an eight-by-eight grid of squares and remove the squares from two opposite corners. Can you cover the remainder of the grid with domino-shaped tiles, each of which exactly covers two adjacent squares? My illustration (Figure 14) shows that you cannot if the eight-by-eight grid is replaced by a four-by-four one. Suppose you decide to place a tile in the position I have marked A. It is easy to see that you are then forced to put tiles in positions B, C, D, and E, leaving a square that cannot be covered. Since the top right-hand corner must be covered somehow, and the only other way of doing it leads to similar problems (by the symmetry of the situation), tiling the whole shape is impossible.

If we replace four by five, then tiling the grid is still impossible, for the simple reason that each tile covers two squares and there are 23 squares to cover – an odd number. However, $8^2 - 2 = 62$ is an even number, so we cannot use this argument for an eight-by-eight grid. On the other hand, if you try to find a proof similar to the one I used for a four-by-four grid, you will soon give up, as the number of

49

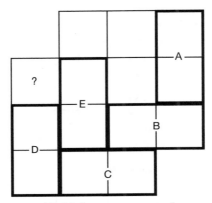

14. Tiling a square grid with the corners removed

possibilities you have to consider is very large. So how should you approach the problem? If you have not come across this question, I would urge you to try to solve it before reading on, or to skip the next paragraph, because if you manage to solve it you will have a good idea of the pleasures of mathematics.

For those who have disregarded my advice, and experience suggests that they will be in the majority, here is one word which is almost the whole proof: chess. A chessboard is an eight-by-eight grid, with its squares coloured alternately black and white (quite unnecessarily as far as the game is concerned, but making it easier to take in visually). Two opposite corner squares will have the same colour. If they are black, as they may as well be, then once they are removed the depleted chessboard has 32 white squares and 30 black ones. Each domino covers exactly one square of each colour, so once you have put down 30 dominoes, you will be left, no matter how you did it, with two white squares, and these you will be unable to cover.

This short argument illustrates very well how a proof can offer more than just a guarantee that a statement is true. For example, we now

have two proofs that the four-by-four grid with two opposite
corners removed cannot be tiled. One is the proof I gave and the
other is the four-by-four version of the chessboard argument.
Both of them establish what we want, but only the second gives
us anything like a *reason* for the tiling being impossible. This
reason instantly tells us that tiling a ten-thousand-by-ten-
thousand grid with two opposite corners removed is also
impossible. By contrast, the first argument tells us only about the
four-by-four case.

It is a notable feature of the second argument that it depends on
a single idea, which, though unexpected, seems very natural as
soon as one has understood it. It often puzzles people when
mathematicians use words like 'elegant', 'beautiful', or even 'witty' to
describe proofs, but an example such as this gives an idea of what
they mean. Music provides a useful analogy: we may be entranced
when a piece moves in an unexpected harmonic direction that
later comes to seem wonderfully appropriate, or when an orchestral
texture appears to be more than the sum of its parts in a way that
we do not fully understand. Mathematical proofs can provide a
similar pleasure with sudden revelations, unexpected yet natural
ideas, and intriguing hints that there is more to be discovered. Of
course, beauty in mathematics is not the same as beauty in music,
but then neither is musical beauty the same as the beauty of a
painting, or a poem, or a human face.

Three obvious-seeming statements that need proofs

An aspect of advanced mathematics that many find puzzling is that
some of its theorems seem too obvious to need proving. Faced with
such a theorem, people will often ask, 'If *that* doesn't count as
obvious, then what does?' A former colleague of mine had a good
answer to this question, which is that a statement is obvious if a
proof instantly springs to mind. In the remainder of this chapter, I
shall give three examples of statements that may appear obvious
but which do not pass this test.

1. The fundamental theorem of arithmetic states that every natural number can be written in one and only one way as a product of prime numbers, give or take the order in which you write them. For example, $36 = 2 \times 2 \times 3 \times 3$, $74 = 2 \times 37$, and 101 is itself a prime number (which is thought of in this context as a 'product' of one prime only). Looking at a few small numbers like this, one rapidly becomes convinced that there will never be two *different* ways of expressing a number as a product of primes. That is the main point of the theorem and it hardly seems to need a proof.

But is it really so obvious? The numbers 7, 13, 19, 37, and 47 are all prime, so if the fundamental theorem of arithmetic is obvious then it should be obvious that $7 \times 13 \times 19$ does not equal 37×47. One can of course check that the two numbers are indeed different (one, as any mathematician will tell you, is more interesting than the other), but that doesn't show that they were *obviously* going to be different, or explain why one could not find two other products of primes, this time giving the same answer. In fact, there is no easy proof of the theorem: if a proof instantly springs to mind, then you have a very unusual mind.

2. Suppose that you tie a slip knot in a normal piece of string and then fuse the ends together, obtaining the shape illustrated in Figure 15, known to mathematicians as the trefoil knot. Is it possible to untie the knot without cutting the string? No, of course it isn't.

Why, though, are we inclined to say 'of course' ? Is there an argument that immediately occurs to us? Perhaps there is – it seems as though any attempt to untie the knot will inevitably make it more tangled rather than less. However, it is difficult to convert this instinct into a valid proof. All that is genuinely obvious is that there is no *simple* way to untie the knot. What is difficult is to rule out the possibility that there is a way of untying the trefoil knot *by making it much more complicated first*. Admittedly, this seems unlikely, but phenomena of this kind do occur in mathematics, and

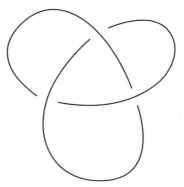

15. A trefoil knot

even in everyday life: for example, in order to tidy a room properly, as opposed to stuffing everything into cupboards, it is often necessary to begin by making it much untidier.

3. A *curve* in the plane means anything that you can draw without lifting your pen off the paper. It is called *simple* if it never crosses itself, and *closed* if it ends up where it began. Figure 16 shows what these definitions mean pictorially. The first curve illustrated, which is both simple and closed, encloses a single region of the plane, which is known as the *interior* of the curve. Clearly, every simple closed curve splits the plane into two parts, the inside and the outside (three parts if one includes the curve itself as a part).

Is this really so clear though? Yes, it certainly is if the curve is not too complicated. But what about the curve shown in Figure 17? If you choose a point somewhere near the middle, it is not altogether obvious whether it lies inside the curve or outside it. Perhaps not, you might say, but there will certainly *be* an inside and an outside, even if the complexity of the curve makes it hard to distinguish them visually.

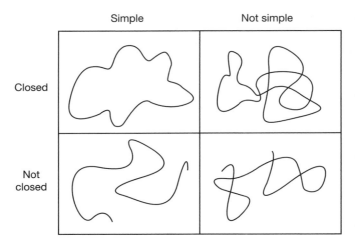

	Simple	Not simple
Closed		
Not closed		

16. Four kinds of curve

How can one justify this conviction? One might attempt to distinguish the inside from the outside as follows. Assuming for a moment that the concepts of inside and outside do make sense, then every time you cross the curve, you must go from the inside to the outside or vice versa. Hence, if you want to decide whether a point P is inside or outside, all you need to do is draw a line that starts at P and ends up at some other point Q that is far enough from the curve to be quite clearly outside. If this line crosses the curve an odd number of times then P is inside; otherwise it is outside.

The trouble with this argument is that it takes various things for granted. For example, how do you know that if you draw *another* line from P, ending at a different point R, you won't get a different answer? (You won't, but this needs to be proved.) The statement that every simple closed curve has an inside and an outside is in fact a famous mathematical theorem, known as the Jordan Curve Theorem. However obvious it may seem, it needs a proof, and all

17. Is the black spot inside the curve or outside it?

known proofs of it are difficult enough to be well beyond the scope of a book like this.

Chapter 4
Limits and infinity

In the last chapter, I tried to indicate how the notion of a mathematical proof can, in principle, be completely formalized. If one starts with certain axioms, follows certain rules, and ends up with an interesting mathematical statement, then that statement will be accepted as a theorem; otherwise, it will not. This idea, of deducing more and more complicated theorems from just a few axioms, goes back to Euclid, who used just five axioms to build up large parts of geometry. (His axioms are discussed in Chapter 6.) Why, one might then ask, did it take until the 20th century for people to realize that this could be done for the whole of mathematics?

The main reason can be summed up in one word: 'infinity'. In one way or another, the concept of infinity is indispensable to mathematics, and yet it is a very hard idea to make rigorous. In this chapter I shall discuss three statements. Each one looks innocent enough at first, but turns out, on closer examination, to involve the infinite. This creates difficulties, and most of this chapter will be about how to deal with them.

1. The square root of 2 is about 1.41421356

Where is infinity involved in a simple statement like the above, which says merely that one smallish number is roughly equal to another? The answer lies in the phrase 'the square root of 2', in which it is implicitly assumed that 2 *has* a square root. If we want to understand the statement completely, this phrase forces us to ask what sort of object the square root of 2 is. And that is where infinity comes in: the square root of 2 is an infinite decimal.

Notice that there is no mention of infinity in the following closely related statement: 1.41421356 squared is close to 2. This statement is entirely finite, and yet it seems to say roughly the same thing. As we shall see later, that is important.

What does it mean to say that there is an infinite decimal which, when squared, gives 2? At school we are taught how to multiply finite decimals but not infinite ones – it is somehow just assumed that they can be added and multiplied. But how is this to be done? To see the sort of difficulty that can arise, let us consider addition first. When we add two finite decimals, such as, say, 2.3859 and 3.1405, we write one under the other and add corresponding digits, starting from the right. We begin by adding the final digits, 9 and 5, together. This gives us 14, so we write down the 4 and carry the 1. Next, we add the penultimate digits, 5 and 0, and the carried 1, obtaining 6. Continuing in this way, we reach the answer, 5.5264.

Now suppose we have two infinite decimals. We cannot start from the right, because an infinite decimal has no last digit. So how can we possibly add them together? There is one obvious answer: start from the left. However, there is a drawback in doing so. If we try it with the finite decimals 2.3859 and 3.1405, for example, we begin by adding the 2 to the 3, obtaining 5. Next, just to the right of the decimal point, we add 3 and 1 and get 4, which is, unfortunately, incorrect.

This incorrectness is inconvenient, but it is not a disaster if we keep our nerve and continue. The next two digits to be added are 8 and 4, and we can respond to them by writing down 2 as the third digit of the answer and *correcting* the second digit by changing it from 4 to 5. This process continues with our writing down 5 as the fourth digit of the answer, which will then be corrected to 6.

Notice that corrections may take place a long time after the digit has been written down. For example, if we add 1.3555555555555555573 to 2.5444444444444444452, then we begin by writing 3.8999999999999999, but that entire string of nines has to be corrected when we get to the next step, which is to add 7 to 5. Then, like a line of dominoes, the nines turn into zeros as we carry one back and back. Nevertheless, the method works, giving an answer of 3.9000000000000000025, and it enables us to give a meaning to the idea of adding two infinite decimals. It is not too hard to see that no digit will ever need to be corrected more than once, so if we have two infinite decimals, then, for example, the 53rd digit of their sum will be what we write down at the 53rd stage of the above process, or the correction of it, should a correction later become necessary.

We would like to make sense of the assertion that there is an infinite decimal whose square is 2. To do this, we must first see how this infinite decimal is generated and then understand what it means to multiply it by itself. As one might expect, multiplication of infinite decimals is more complicated than addition.

First, though, here is a natural way to generate the decimal. It has to lie between 1 and 2, because $1^2 = 1$, which is less than 2, and $2^2 = 4$, which is greater. If you work out 1.1^2, 1.2^2, 1.3^2, and so on up to 1.9^2 you find that $1.4^2 = 1.96$, which is less than 2, and $1.5^2 = 2.25$, which is greater. So $\sqrt{2}$ must lie between 1.4 and 1.5, and therefore its decimal expansion must begin 1.4. Now suppose that you have worked out in this way that the first eight digits of $\sqrt{2}$ are 1.4142135. You can then do the following calculations, which show that the next digit is 6.

$$1.41421350^2 = 1.9999998236822500$$
$$1.41421351^2 = 1.9999998519665201$$
$$1.41421352^2 = 1.9999998802507904$$
$$1.41421353^2 = 1.9999999085350609$$
$$1.41421354^2 = 1.9999999368193316$$
$$1.41421355^2 = 1.9999999651036025$$
$$1.41421356^2 = 1.9999999933878736$$
$$1.41421357^2 = 2.0000000216721449$$

Repeating this procedure, you can generate as many digits as you like. Though you will never actually finish, you do at least have an unambiguous way of defining the nth digit after the decimal point, whatever the value of n: it will be the same as the final digit of the largest decimal that squares to less than 2 and has n digits after the decimal point. For example, 1.41 is the largest decimal that squares to less than 2 with two digits after the decimal point, so the square root of two begins 1.41.

Let us call the resulting infinite decimal x. What makes us so confident that $x^2 = 2$? We might argue as follows.

$$1^2 = 1$$
$$1.4^2 = 1.96$$
$$1.41^2 = 1.9881$$
$$1.414^2 = 1.999396$$
$$1.4142^2 = 1.99996164$$
$$1.41421^2 = 1.9999899241$$
$$1.414213^2 = 1.999998409469$$
$$1.4142135^2 = 1.99999982368225$$
$$1.41421356^2 = 1.9999999933878736$$

As the above table of calculations demonstrates, the more digits we use of the decimal expansion of $\sqrt{2}$, the more nines we get after the decimal point when we multiply the number by itself. Therefore, if we use the entire infinite expansion of $\sqrt{2}$, we should get infinitely many nines, and 1.99999999 . . . (one point nine recurring) equals 2.

This argument leads to two difficulties. First, why does one point nine recurring equal two? Second, and more serious, what does it mean to 'use the entire infinite expansion'? That is what we were trying to understand in the first place.

To dispose of the first objection, we must once again set aside any Platonic instincts. It is an accepted truth of mathematics that one point nine recurring equals two, but this truth was not discovered by some process of metaphysical reasoning. Rather, it is a *convention*. However, it is by no means an arbitrary convention, because not adopting it forces one either to invent strange new objects or to abandon some of the familiar rules of arithmetic. For example, if you hold that 1.999999 . . . does not equal 2, then what is $2 - 1.999999 . . .$? If it is zero, then you have abandoned the useful rule that x must equal y whenever $x - y = 0$. If it is not zero, then it does not have a conventional decimal expansion (otherwise, subtract it from two and you will not get one point nine recurring but something smaller) so you are forced to invent a new object such as 'nought followed by a point, then infinitely many noughts, and *then* a one'. If you do this, then your difficulties are only just beginning. What do you get when you multiply this mysterious number by itself? Infinitely many noughts, then infinitely many noughts again, and *then* a one? What happens if you multiply it by ten instead? Do you get 'infinity minus one' noughts followed by a one? What is the decimal expansion of 1/3? Now multiply that number by 3. Is the answer 1 or 0.999999 . . . ? If you follow the usual convention, then tricky questions of this kind do not arise. (Tricky but not impossible: a coherent notion of 'infinitesimal' numbers was discovered by Abraham Robinson in the 1960s, but non-standard analysis, as his theory is called, has not become part of the mathematical mainstream.)

The second difficulty is a more genuine one, but it can be circumvented. Instead of trying to imagine what would actually happen if one applied some kind of long multiplication procedure to infinite decimals, one interprets the statement $x^2 = 2$ as meaning simply that the more digits one takes of x, the closer the square of the resulting number is to 2, just as we observed. To be more precise, suppose you insist that you want a number that, when squared, produces a number that begins 1.9999. . . . I will suggest the number 1.41421, given by the first few digits of x. Since 1.41421 is very close to 1.41422, I expect that their squares are also very close (and this can be proved quite easily). But because of how we chose x, 1.41421^2 is less than 2 and 1.41422^2 is greater than 2. It follows that both numbers are very close to 2. Just to check: $1.41421^2 = 1.9999899241$, so I have found a number with the property you wanted. If you now ask for a number that, when squared, begins

1.99999999999999999999999999999999999999999999999999 . . .,

I can use exactly the same argument, but with a few more digits of x. (It turns out that if you want n nines then $n + 1$ digits after the decimal point will always be enough.) The fact that I can do this, however many nines you want, is what is meant by saying that the infinite decimal x, when multiplied by itself, equals 2.

Notice that what we have done is to 'tame' the infinite, by interpreting a statement that involves infinity as nothing more than a colourful shorthand for a more cumbersome statement that doesn't. The neat infinite statement is 'x is an infinite decimal that squares to 2'. The translation is something like, 'There is a rule that, for any n, unambiguously determines the nth digit of x. This allows us to form arbitrarily long finite decimals, and their squares can be made as close as we like to 2 simply by choosing them long enough.'

Am I saying that the true meaning of the apparently simple statement that $x^2 = 2$ is actually very complicated? In a way I am – the statement really does have hidden complexities – but in a more important way I am not. It is hard work to define addition and multiplication of infinite decimals without mentioning infinity, and one must check that the resulting complicated definitions obey the rules set out in Chapter 2, such as the commutative and associative laws. However, once this has been done, we are free to think abstractly again. What matters about x is that it squares to two. What matters about the word 'squares' is that its meaning is based on *some* definition of multiplication that obeys the appropriate rules. It doesn't really matter what the trillionth digit of x is and it doesn't really matter that the definition of multiplication is somewhat complicated.

2. We reached a speed of 40 m.p.h. just as we passed that lamp-post

Suppose you are in an accelerating car and you have watched the speedometer move steadily from 30 m.p.h. to 50 m.p.h. It is tempting to say that just for an instant – the exact instant at which the arm of the speedometer passed 40 – the car was travelling at 40 m.p.h. Before that instant it was slower and afterwards it was faster. But what does it mean to say that the speed of a car is 40 m.p.h. just for an instant? If a car is not accelerating, then we can measure how many miles it goes in an hour, and that gives us its speed. (Alternatively, and more practically, we can see how far it goes in 30 seconds and multiply by 120.) However, this method obviously does not work with an accelerating car: if one measures how far it has gone in a certain time, all one can calculate is the *average* speed during that time, which does not tell us the speed at any given moment.

The problem would disappear if we could measure how far the car travelled during an *infinitely small* period of time, because then the speed would not have time to change. If the period of time lasted for t hours, where t was some infinitely small number, then we would measure how many miles the car travelled during those t hours, take our answer s, which would also be infinitely small of course, and divide it by t to obtain the instantaneous speed of the car.

This ridiculous fantasy leads to problems very similar to those encountered when we briefly entertained the idea that one point nine recurring might not equal two. Is t zero? If so, then quite clearly s must be as well (a car cannot travel any distance in no time at all). But one cannot divide zero by zero and obtain an unambiguous answer. On the other hand, if t is not zero, then the car accelerates during those t hours and the measurement is invalid.

The way to understand instantaneous speed is to exploit the fact that the car does not have time to accelerate *very much* if t is very small – say a hundredth of a second. Suppose for a moment that we do not try to calculate the speed exactly, but settle instead for a good estimate. Then, if our measuring devices are accurate, we can see how far the car goes in a hundredth of a second, and multiply that distance by the number of hundredths of a second in an hour, or 360,000. The answer will not be quite right, but since the car cannot accelerate much in a hundredth of a second it will give us a close approximation.

This situation is reminiscent of the fact that 1.4142135^2 is a close approximation to 2, and it allows us to avoid worrying about the infinite, or in this case infinitely small, in a very similar way. Suppose that instead of measuring how far the car went in a hundredth of a second we had measured its distance over a millionth of a second. The car would have accelerated even less during that time, so our answer would have been more accurate still. This observation gives us a way of translating the statement, 'The car is travelling at 40 m.p.h. . . . NOW!' into the following more complicated finite one: 'If you specify what margin of error I am allowed, then as long as t is a small enough number of hours (typically far less than one) I can see how many miles the car travels in t hours, divide by t and obtain an answer that will be at least as close to 40 m.p.h. as the error margin you allowed me.' For example, if t is small enough, then I can guarantee that my estimate will be between 39.99 and 40.01. If you have asked for an answer accurate to within 0.0001, then I may have to make t smaller, but as long as it is small enough I can give you the accuracy you want.

Once again, we are regarding a statement that involves infinity as a convenient way of expressing a more complicated statement concerning approximations. Another word, which can be more suggestive, is 'limit'. An infinite decimal is a limit of a sequence of finite decimals, and the instantaneous speed is the limit of the estimates one makes by measuring the distance travelled over shorter and shorter periods of time. Mathematicians often talk about what happens 'in the limit', or 'at infinity', but they are aware as they do so that they are not being wholly serious. If pressed to say exactly what they mean, they will start to talk about approximations instead.

3. The area of a circle of radius r is πr^2

The realization that the infinite can be understood in terms of the finite was one of the great triumphs of 19th-century mathematics, although its roots go back much earlier. In discussing my next example, how to calculate the area of a circle, I shall use an argument invented by Archimedes in the 3rd century BC. Before we do this calculation, though, we ought to decide what it is that we are calculating, and this is not as easy as one might think. What *is* area? Of course, it is something like the *amount of stuff* in the shape (two-dimensional stuff, that is), but how can one make this precise?

Whatever it is, it certainly seems to be easy to calculate for certain shapes. For example, if a rectangle has side lengths a and b, then its area is ab. Any right-angled triangle can be thought of as the result of cutting a rectangle in half down one of its diagonals, so its area is half that of the corresponding rectangle. Any triangle can be cut into two right-angled triangles, and any polygon can be split up into triangles. Therefore, it is not too difficult to work out the area of a polygon. Instead of worrying about what exactly it is that we have calculated, we can simply *define* the area of a polygon to be the result of the calculation (once we have convinced ourselves that chopping a polygon into triangles in two different ways will not give us two different answers).

Our problems begin when we start to look at shapes with curved boundaries. It is not possible to cut a circle up into a few triangles. So what are we talking about when we say that its area is πr^2?

This is another example where the abstract approach is very helpful. Let us concentrate not on what area *is*, but on what it *does*. This suggestion needs to be clarified since area doesn't seem to do all that much – surely it is just there. What I mean is that we should focus on the properties that any reasonable concept of area will have. Here are five.

Ar1 If you slide a shape about, its area does not change. (Or more formally: two congruent shapes have the same area.)

Ar2 If one shape is entirely contained in another, then the area of the first is no larger than the area of the second.

Ar3 The area of a rectangle is obtained by multiplying its two side lengths.

Ar4 If you cut a shape into a few pieces, then the areas of the pieces add up to the area of the original shape.

Ar5 If you expand a shape by a factor of 2 in every direction, then its area multiplies by 4.

If you look back, you will see that we used properties **Ar1**, **Ar3**, and **Ar4** to calculate the area of a right-angled triangle. Property **Ar2** may seem so obvious as to be hardly worth mentioning, but this is what one expects of axioms, and we shall see later that it is very useful. Property **Ar5**, though important, is not actually needed as an axiom because it can be deduced from the others.

How can we use these properties to say what we mean by the area of a circle? The message of this chapter so far is that it may be fruitful to think about *approximating* the area, rather than defining it in one go. This can be done quite easily as follows. Imagine that a shape is drawn on a piece of graph paper with a very fine grid of squares. We know the area of these squares, by property **Ar3** (since a square is a special kind of rectangle), so we could try estimating the area of the shape by counting how many of the squares lie completely within it. If, for example, the shape contains 144 squares, then the area of the shape is at least 144 times the area of each square. Notice that what we have really calculated is the area of a shape made out of the 144 squares, which is easily determined by properties **Ar3** and **Ar4**.

For the shape illustrated in Figure 18 this does not give the right answer because there are several squares that are partly in the shape

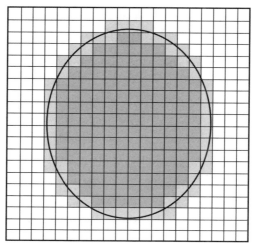

18. **Approximating the area of a curved shape**

and partly out of it, so we have not taken account of all of the area. However, there is an obvious way to improve the estimate, which is to divide each square into four smaller ones and use those instead. As before, some of the squares will be partly in and partly out, but we will now have included a little more of the shape amongst the squares that are completely inside it. In general, the finer the grid of squares, the more of the original shape we take account of in our calculation. We find (and this is not quite as obvious as it seems) that as we take finer and finer grids, with smaller and smaller squares, the results of our calculations are closer and closer to some number, just as the results of squaring better and better approximations to $\sqrt{2}$ become closer and closer to 2, and we *define* this number to be the area of the shape.

Thus, to the mathematically inclined, the statement that some shape has an area of one square yard means the following. If a certain amount of error is regarded as tolerable, then, however small it is, one can choose a sufficiently fine grid of squares, do the approximate calculation by adding up the areas of the squares inside the shape, and arrive at an answer which will differ from one square yard by less than that amount. (Somewhere in the back of one's mind, but firmly suppressed, may be the idea that 'in the limit' one could use infinitely many infinitely small squares and get the answer exactly right.)

Another way of putting it, which is perhaps clearer, is this. If a curved shape has an area of exactly 12 square centimetres, and I am required to demonstrate this using a grid of squares, then my task is impossible – I would need infinitely many of them. If, however, you give me any number *other* than 12, such as 11.9, say, then I can use a grid of squares to prove conclusively that the area of the shape is *not* that number: all I have to do is choose a grid fine enough for the area left out to be less than 0.1 square centimetres. In other words, I can do without infinity if, instead of proving that the area is 12, I settle for proving that it isn't anything else. The area of the shape is the one number I cannot disprove.

These ideas give a satisfactory definition of area, but they still leave us with a problem. How can we show that if we use the above procedure to estimate the area of a circle of radius r, then our estimates will get closer and closer to πr^2? The answer for most shapes is that one has to use the integral calculus, which I do not discuss in this book, but for the circle one can use an ingenious argument of Archimedes, as I mentioned earlier.

Figure 19 shows a circle cut into slices and an approximately rectangular shape formed by taking the slices apart and reassembling them. Because the slices are thin, the height of the rectangle is approximately the radius, r, of the circle. Again because the slices are thin, the top and bottom sides of the approximate rectangle are approximately straight lines. Since each of these sides uses half the circumference of the circle, and by the definition of π the circumference has length $2\pi r$, their lengths are approximately πr. Therefore, the area of the approximate rectangle is $r \times \pi r = \pi r^2$ – at least approximately.

Of course, it is in fact exactly πr^2 since all we have done is cut up a circle and move the pieces around, but we do not yet know this. The argument so far may have already convinced you, but it is not quite finished, as we must establish that the above approximation becomes closer and closer to πr^2 as the number of slices becomes larger and larger. Very briefly, one way to do this is to take two regular polygons, one just contained in the circle and the other just

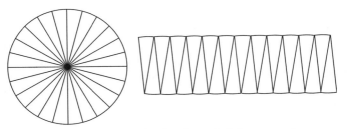

19. Archimedes' method for showing that a circle has area πr^2

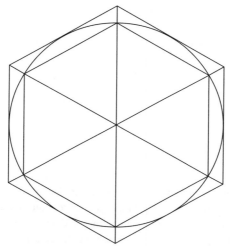

20. Approximating a circle by a polygon

containing it. Figure 20 illustrates this with hexagons. The perimeter of the inner polygon is shorter than the circumference of the circle, while that of the outer polygon is longer. The two polygons can each be cut into triangular slices and reassembled into parallelograms. A straightforward calculation shows that the smaller parallelogram has an area of less than r times half the perimeter of the inner polygon, and hence less than πr^2. Similarly, the area of the larger polygon is greater than πr^2. And yet, if the number of slices is large enough, the difference in area between the two polygons can be made as small as one likes. Since the circle always contains the smaller one and is contained in the larger, its area must be exactly πr^2.

Chapter 5
Dimension

A notable feature of advanced mathematics is that much of it is concerned with geometry in more than three dimensions. This fact is baffling to non-mathematicians: lines and curves are one-dimensional, surfaces are two-dimensional, and solid objects are three-dimensional, but how could something be four-dimensional? Once an object has height, width, and depth, it completely fills up a portion of space, and there just doesn't seem to be scope for any further dimensions. It is sometimes suggested that the fourth dimension is time, which is a good answer in certain contexts, such as special relativity, but does not help us to make sense of, say, twenty-six-dimensional or even infinite-dimensional geometry, both of which are of mathematical importance.

High-dimensional geometry is yet another example of a concept that is best understood from an abstract point of view. Rather than worrying about the *existence*, or otherwise, of twenty-six-dimensional space, let us think about its *properties*. You might wonder how it is possible to consider the properties of something without establishing first that it exists, but this worry is easily disposed of. If you leave out the words 'of something', then the question becomes: how can one consider a set of properties without first establishing that there is something that *has* those properties? But this is not difficult at all. For example, one can speculate about the character a female president of the United States would be likely to have, even though there is no guarantee that there will ever be one.

What sort of properties might we expect of twenty-six dimensional space? The most obvious one, the property that would *make* it twenty-six dimensional, is that it should take twenty-six numbers to specify a point, just as it takes two numbers in two dimensions and three in three. Another is that if you take a twenty-six dimensional shape and expand it by a factor of two in every direction, then its 'volume', assuming we can make sense of it, should multiply by 2^{26}. And so on.

Such speculations would not be very interesting if it turned out that there was something logically inconsistent about the very notion of twenty-six dimensional space. To reassure ourselves about this, we would after all like to show that it exists – which it obviously couldn't if it involved an inconsistency – but in a mathematical rather than a physical sense. What this means is that we need to define an appropriate model. It may not necessarily be a model *of* anything, but if it has all the properties we expect, it will show that those properties are consistent. As so often happens, though, the model we shall define turns out to be very useful.

How to define high-dimensional space

Defining the model is surprisingly easy to do, as soon as one has had one idea: coordinates. As I have said, a point in two dimensions can be specified using two numbers, while a point in three dimensions needs three. The usual way to do this is with Cartesian coordinates, so called because they were invented by Descartes (who maintained that he had been given the idea in a dream). In two dimensions you start with two directions at right angles to each other. For example, one might be to the right and the other directly upwards, as shown in Figure 21. Given any point in the plane, you can reach it by going a certain distance horizontally (if you go left then you regard yourself as having gone a negative distance to the right), then

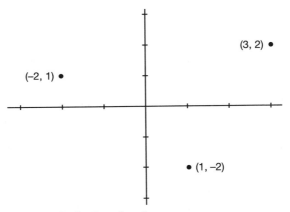

21. Three points in the Cartesian plane

turning through ninety degrees and going some other distance vertically. These two distances give you two numbers, and these numbers are the coordinates of the point you have reached. Figure 21 shows the points with coordinates (3, 2) (three to the right and two up), (−2, 1) (two to the left and one up), and (1, −2) (one to the right and two down). Exactly the same procedure works in three dimensions, that is, in space, except that you must use three directions, such as forwards, to the right, and upwards.

Now let us change our point of view very slightly. Instead of calling the two (or three) numbers *the coordinates of* a point in space, let us say that the numbers *are* the point. That is, instead of saying 'the point with coordinates (5, 3)', let us say 'the point (5, 3)'. One might regard doing this as a mere linguistic convenience, but actually it is more than that. It is replacing real, physical space with a mathematical model of space. Our mathematical model of two-dimensional space consists of pairs of real numbers (a, b). Although these pairs of numbers are not themselves points in space, we call them points because we wish to remind ourselves that that is what they represent. Similarly, we can obtain a model of three-dimensional space by taking all triples (a, b, c), and again calling them points. There is now an obvious way of defining points in, say, eight-dimensional space. They are nothing other than octuples of real numbers. For example, here are two points: $(1, 3, -1, 4, 0, 0, 6, 7)$ and $(5, \pi, -3/2, \sqrt{2}, 17, 89.93, -12, \sqrt{2} + 1)$.

I have now defined a mathematical model of sorts, but it is not yet worthy of being called a model of eight-dimensional *space*, because the word 'space' carries with it many geometrical connotations which I have not yet described in terms of the model: there is more to space than just a vast collection of single points. For example, we talk about the distance between a pair of points, and about straight lines, circles, and other geometrical shapes. What are the counterparts of these ideas in higher dimensions?

There is a general method for answering many questions of this kind. Given a familiar concept from two or three dimensions, first describe it entirely in terms of coordinates and then hope that the

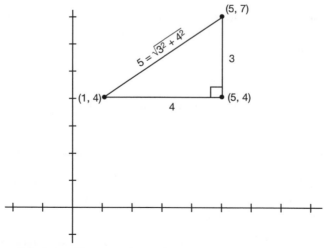

22. Calculating distances using Pythagoras' theorem

generalization to higher dimensions becomes obvious. Let us see how this works for the concept of distance.

Given two points in the plane, such as (1, 4) and (5, 7), we can calculate their distance as follows. We begin by forming a right-angled triangle with the extra point (5, 4), as illustrated in Figure 22. We then notice that the line joining the points (1, 4) to (5, 7) is the hypotenuse of this triangle, which means that its length can be calculated using Pythagoras' theorem. The lengths of the other two sides are $5 - 1 = 4$ and $7 - 4 = 3$, so the length of the hypotenuse is $\sqrt{4^2 + 3^2} = \sqrt{16 + 9} = 5$. Thus, the distance between the two points is 5. Applying this method to a general pair of points (a, b) and (c, d), we obtain a right-angled triangle with these two points at either end of the hypotenuse and two other side lengths of $|c - a|$ (this means the difference between c and a) and $|d - b|$. Pythagoras' theorem then tells us that the distance between the two points is given by the formula

$$\sqrt{(c - a)^2 + (d - b)^2}$$

A similar but slightly more complicated argument works in three dimensions and shows that the distance between two points (a, b, c) and (d, e, f) is

$$\sqrt{(d - a)^2 + (e - b)^2 + (f - c)^2}$$

In other words, to calculate the distance between two points, you add up the squares of the differences between corresponding coordinates and then take the square root. (Briefly, the justification is this. The triangle T with vertices (a, b, c), (a, b, f), and (d, e, f) has a right angle at (a, b, f). The distance from (a, b, c) to (a, b, f) is $|f - c|$ and the distance from (a, b, f) to (d, e, f) is, by the two-dimensional formula, $\sqrt{(d - a)^2 + (e - b)^2}$. The result now follows from Pythagoras' theorem applied to T.)

An interesting feature of this statement is that it makes no mention of the fact that the points were supposed to be three-dimensional.

We have therefore stumbled on a method for calculating distances in *any* number of dimensions. For example, the distance between the two points (1, 0, –1, 4, 2) and (3, 1, 1, 1, –1) (which lie in five-dimensional space) is

$$\sqrt{(3-1)^2 + (1-0)^2 + (1-(-1))^2 + (1-4)^2 + (-1-2)^2} =$$
$$\sqrt{4 + 1 + 4 + 9 + 9} = \sqrt{27}$$

Now this way of putting things is a little misleading because it suggests that there was always a distance between any pair of five-dimensional points (remember that a five-dimensional point means nothing more than a sequence of five real numbers) and that we have discovered how to work these distances out. Actually, however, what we have done is *define* a notion of distance. No physical reality forces us to decide that five-dimensional distance should be calculated in the manner described. On the other hand, this method is so clearly the natural generalization of what we do in two and three dimensions that it would be strange to adopt any other definition.

Once distance has been defined, we can begin to generalize other concepts. For example, a sphere is clearly the three-dimensional equivalent of a circle. What would a four-dimensional 'sphere' be? As with distance, we can answer this question if we can describe the two- and three-dimensional versions in a way that does not actually mention the number of dimensions. This is not at all hard: circles and spheres can both be described as the set of all points at a fixed distance (the radius) from some given point (the centre). There is nothing to stop us from using exactly the same definition for four-dimensional spheres, or eighty-seven-dimensional spheres for that matter. For example, the four-dimensional sphere of radius 3 about the point (1, 1, 0, 0) is the set of all (four-dimensional) points at distance 3 from (1, 1, 0, 0). A four-dimensional point is a sequence (a, b, c, d) of four real numbers. Its distance from (1, 1, 0, 0) is (according to our earlier definition)

$$\sqrt{(a-1)^2 + (b-1)^2 + c^2 + d^2}$$

Therefore, another description of this four-dimensional sphere is that it is the set of all quadruples (a, b, c, d) for which

$$\sqrt{(a-1)^2 + (b-1)^2 + c^2 + d^2} = 3$$

For example, $(1, -1, 2, 1)$ is such a quadruple, so this is a point in the given four-dimensional sphere.

Another concept that can be generalized is that of a square in two dimensions and a cube in three. As Figure 23 makes clear, the set of all points (a, b) such that both a and b lie between 0 and 1 forms a square of side length 1, and its four vertices are the points $(0, 0)$, $(0, 1)$, $(1, 0)$, and $(1, 1)$. In three dimensions one can define a cube by taking all points (a, b, c) such that a, b, and c all lie between 0 and 1. Now there are eight vertices: $(0, 0, 0)$, $(0, 0, 1)$, $(0, 1, 0)$, $(0, 1, 1)$, $(1, 0, 0)$, $(1, 0, 1)$, $(1, 1, 0)$, and $(1, 1, 1)$. Similar definitions are obviously possible in higher dimensions. For example, one can obtain a six-dimensional cube, or rather a mathematical construction clearly worthy of the name, by taking all points (a, b, c, d, e, f) with all their coordinates lying between 0 and 1. The vertices will be all points for which every coordinate is 0 or 1: it is not hard to see that the number of vertices doubles each time you add a dimension, so in this case there are 64 of them.

One can do much more than simply *define* shapes. Let me illustrate this briefly by calculating the number of edges of a five-dimensional cube. It is not immediately obvious what an edge is, but for this we can take our cue from what happens in two and three dimensions: an edge is the line that joins two neighbouring vertices, and two vertices are regarded as neighbours if they differ in precisely one coordinate. A typical vertex in the five-dimensional cube is a point such as $(0, 0, 1, 0, 1)$ and, according to the definition just given, its neighbours are $(1, 0, 1, 0, 1)$, $(0, 1, 1, 0, 1)$, $(0, 0, 0, 0, 1)$, $(0, 0, 1, 1, 1)$, and $(0, 0, 1, 0, 0)$. In general, each vertex has five neighbours, and hence five edges coming out of it. (I leave it to the reader to generalize from two and three dimensions the notion of the line joining two neighbouring vertices. For this calculation it does not matter.) Since there are $2^5 = 32$ vertices, it looks as though there are $32 \times 5 = 160$ edges. However, we have counted each edge twice – once each for its two end points – so the correct answer is half of 160, that is, 80.

One way of summarizing what we are doing is to say that we are converting geometry into algebra, using coordinates to translate geometrical concepts into equivalent concepts that involve only relationships between numbers. Although we cannot directly generalize the geometry, we *can* generalize the algebra, and it seems reasonable to call this generalization higher-dimensional geometry. Obviously, five-dimensional geometry is not as directly related to our immediate experience as three-dimensional geometry, but this does not make it impossible to think about, or prevent it from being useful as a model.

Can four-dimensional space be visualized?

In fact, the seemingly obvious statement that it is possible to visualize three-dimensional objects but not four-dimensional ones does not really stand up to close scrutiny. Although visualizing an object feels rather like looking at it, there are important differences between the two experiences. For example, if I am asked to visualize a room with which I am familiar, but not very familiar, I have no difficulty in doing so. If I am then asked simple questions about it, such as how many chairs it contains or what colour the floor is, I am often unable to answer them. This shows that, whatever a mental image is, it is not a photographic representation.

In a mathematical context, the important difference between being able to visualize something and not being able to is that in the former case one can somehow answer questions directly rather than having to stop and calculate. This directness is of course a matter of degree, but it is no less real for that. For example, if I am asked to give the number of edges of a three-dimensional cube, I can do it by 'just seeing' that there are four edges round the top, four round the bottom, and four going from the top to the bottom, making twelve in all.

In higher dimensions, 'just seeing' becomes more difficult, and one is often forced to argue more as I did when discussing the analogous question in five dimensions. However, it is sometimes possible. For example, I can think of a four-dimensional cube as consisting of two three-dimensional cubes facing each other, with corresponding vertices joined by edges (in the fourth dimension), just as a three-dimensional cube consists of two squares facing each other with corresponding vertices joined. Although I do not have a completely clear picture of four-dimensional space, I can still 'see' that there are twelve edges for each of the two three-dimensional cubes, and eight edges linking their vertices together. This gives a total of $12 + 12 + 8 = 32$. Then I can 'just see' that a five-dimensional cube is made up of two of these, again with corresponding vertices linked, making a total of $32 + 32 + 16 = 80$ edges (32 for each four-dimensional cube and 16 for the edges between them), exactly the answer I obtained before. Thus, I have some rudimentary ability to visualize in four and five dimensions. (If you are bothered by the word 'visualize' then you can use another one, such as 'conceptualize'.) Of course, it is much harder than visualizing in three dimensions – for example, I cannot directly answer questions about what happens when you rotate a four-dimensional cube, whereas I can for a three-dimensional one – but it is also distinctly easier than fifty-three-dimensional visualization, which it could not be if they were both impossible. Some mathematicians specialize in four-dimensional geometry, and their powers of four-dimensional visualization are highly developed.

This psychological point has an importance to mathematics that goes well beyond geometry. One of the pleasures of devoting one's

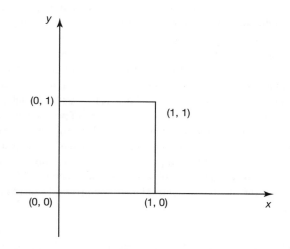

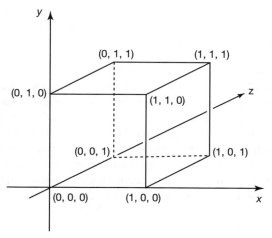

23. A unit square and a unit cube

life to mathematical research is that, as one gains in expertise, one finds that one can 'just see' answers to more and more questions that might once have required an hour or two of hard thought, and the questions do not have to be geometrical. An elementary example of this is the statement that $471 \times 638 = 638 \times 471$. One could verify this by doing two different long multiplications and checking that they give the same answer. However, if one thinks instead of a grid of points arranged in a 471-by-638 rectangle, one can see that the first sum adds up the number of points in each row and the second the number of points in each column, and these must of course give the same answer. Notice that a mental picture here is quite different from a photograph: did you really visualize a 471-by-638 rectangle rather than a 463-by-641 rectangle? Could you count the number of points along the shorter side, just to check?

What is the point of higher-dimensional geometry?

It is one thing to demonstrate that some sense can be made of the idea of higher-dimensional geometry, but quite another to explain why it is a subject worth taking seriously. Earlier in the chapter, I claimed that it was useful as a model, but how can this be, given that the actual space we inhabit is three-dimensional?

The answer to this question is rather simple. One of the points I made in Chapter 1 was that a model can have many different uses. Even two- and three-dimensional geometry are used for many purposes other than a straightforward modelling of physical space. For example, we often represent the motion of an object by drawing a graph that records the distance it has travelled at different times. This graph will be a curve in the plane, and geometrical properties of the curve correspond to information about the motion of the object. Why is two-dimensional geometry appropriate for modelling this motion? Because there are two numbers of interest – the time elapsed and the distance travelled – and, as I have said, one can think of two-dimensional space as the collection of all pairs of numbers.

This gives us a clue about why higher-dimensional geometry can be useful. There may not be any high-dimensional space lurking in the universe, but there are plenty of situations in which we need to consider collections of several numbers. I shall very briefly describe two, after which it should become obvious that there are many more.

Suppose that I wish to describe the position of a chair. If it is standing upright, then its position is completely determined by the points where two of its legs meet the floor. These two points can each be described by two coordinates. The result is that four numbers can be used to describe the position of the chair. However, these four numbers are related, because the distance between the bottoms of the legs is fixed. If this distance is d and the legs meet the floor at the points (p, q) and (r, s) then $(p - r)^2 + (q - s)^2 = d^2$, by Pythagoras' theorem. This puts a constraint on p, q, r, and s, and one way to describe this constraint uses geometrical language: the point (p, q, r, s), which belongs to four-dimensional space, is forced to lie in a certain three-dimensional 'surface'. More complicated physical systems can be analysed in a similar way, and the dimensions become much higher.

Multi-dimensional geometry is also very important in economics. If, for example, you are wondering whether it is wise to buy shares in a company, then much of the information that will help you make your decision comes in the form of numbers – the size of the workforce, the values of various assets, the costs of raw materials, the rate of interest, and so on. These numbers, taken as a sequence, can be thought of as a point in some high-dimensional space. What you would like to do, probably by analysing many other similar companies, is identify a region of that space, the region where it is a good idea to buy the shares.

Fractional dimension

If one thing seems obvious from the discussion so far, it is that the dimension of any shape will always be an integer. What could it possibly mean to say that you need two and a half coordinates to specify a point – even a mathematical one?

This argument may look compelling, but we were faced with a very similar difficulty before we defined the number $2^{3/2}$ in Chapter 2, and managed to circumvent it by using the abstract method. Can we do something similar for dimension? If we want to, then we must find some property associated with dimension that does not immediately imply that it is a whole number. This rules out anything to do with the number of coordinates, which seems so closely bound up with the very idea of dimension that it is hard to think of anything else. There is, however, another property, mentioned briefly at the beginning of this chapter, that gives us exactly what we need.

An important aspect of geometry which varies with dimension is the rule that determines what happens to the size of a shape when you expand it by a factor of t in every direction. By size, I mean length, area, or volume. In one dimension, the size multiplies by t, or t^1, in two dimensions it multiplies by t^2, and in three it multiplies by t^3. Thus, the power of t tells us the dimension of the shape.

So far we have not quite managed to banish whole numbers from the picture, because the numbers two and three are implicit in the words 'area' and 'volume'. However, we can do without these words as follows. Why is it that a square of side length three has nine times the area of a square of side length one? The reason is that one can divide the larger square into nine congruent copies of the smaller one (see Figure 24.). Similarly, a three-by-three-by-three cube can be divided into twenty-seven one-by-one-by-one cubes, so its volume is twenty-seven times that of a one-by-one-by-one cube. So we can say that a cube is three-dimensional because if it is expanded

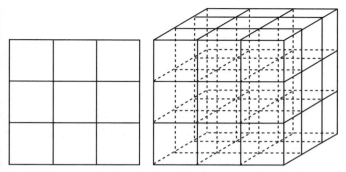

24. Dividing a square into $9 = 3^2$ smaller squares, and a cube into $27 = 3^3$ smaller cubes

by a factor of t, where t is a whole number greater than 1, then the new cube can be divided into t^3 copies of the old one. Note that the word 'volume' did not appear in the last sentence.

We may now ask: is there a shape for which we can reason as above and obtain an answer that is not an integer? The answer is yes. One of the simplest examples is known as the Koch snowflake. It is not possible to describe it directly: instead, it is defined as the limit of the following process. Start with a straight line segment, of length one, say. Then divide it into three equal pieces, and replace the middle piece with the other two sides of the equilateral triangle that has this middle piece as its base. The result is a shape built out of four straight line segments, each of length one-third. Divide each of these into three equal pieces, and again replace each middle piece by the other two sides of an equilateral triangle. Now the result is a shape made of sixteen line segments, each of length one-ninth. It is clear how to continue this process: the first few stages are illustrated in Figure 25. It is not too hard to prove rigorously that this process leads to a limiting shape, as the pictures suggest, and this shape is the Koch snowflake. (It looks more like a snowflake if you take three copies of it and put them together round a triangle.)

25. Building up the Koch snowflake

The Koch snowflake has several interesting features. The one that will concern us is that it can be built out of smaller copies of itself. Once again, this can be seen from the picture: it consists of four copies, and each copy is a shrinking of the whole shape by a factor of one-third. Let us now consider what that tells us about its dimension.

If a shape is d-dimensional, then when we shrink it by a factor of one-third its size should go down by a factor of 3^d. (As we have seen, this is true when d is 1, 2, or 3.) Thus, if we can build it out of smaller copies, then we should need 3^d of them. Since four copies are needed for the Koch snowflake, its dimension d should be the number for which $3^d = 4$. Since $3^1 = 3$ and $3^2 = 9$, this means that d lies between 1 and 2, and so is not a whole number. In fact, it is $\log_3 4$, which is approximately 1.2618595.

This calculation depends on the fact that the Koch snowflake can be decomposed into smaller copies of itself, which is a very unusual feature: even a circle doesn't have it. However, it is possible to develop the above idea and give a definition of dimension that is much more widely applicable. As with our other uses of the abstract method, this does not mean that we have discovered the 'true dimension' of the Koch snowflake and similar exotic shapes – but merely that we have found the only possible definition consistent with certain properties. In fact, there are other ways of defining dimension that give different answers. For example, the Koch snowflake has a 'topological dimension' of 1. Roughly speaking, this is because, like a line, it can be broken into two disconnected parts by the removal of any one of its interior points.

This sheds interesting light on the twin processes of abstraction and generalization. I have suggested that to generalize a concept one should find some properties associated with it and generalize those. Often there is only one natural way to do this, but sometimes different sets of properties lead to different generalizations, and sometimes more than one generalization is fruitful.

Chapter 6
Geometry

Perhaps the most influential mathematics book of all time is Euclid's *Elements*, written in around 300 BC. Although Euclid lived over two thousand years ago, he was in many ways the first recognizably modern mathematician – or at least the first that we know about. In particular, he was the first author to make systematic use of the axiomatic method, beginning his book with five axioms and deducing from them a large body of geometrical theorems. The geometry with which most people are familiar, if they are familiar with any at all, is the geometry of Euclid, but at the research level the word 'geometry' has a much broader definition: today's geometers do not spend much of their time with a ruler and compass.

Euclidean geometry

Here are Euclid's axioms. I follow the normal convention and use the word 'line' for a line that extends indefinitely in both directions. A 'line segment' will mean a line with two end-points.

1. Any two points can be joined by exactly one line segment.
2. Any line segment can be extended to exactly one line.
3. Given any point P and any length r, there is a circle of radius r with P as its centre.
4. Any two right angles are congruent.

5. If a straight line N intersects two straight lines L and M, and if the interior angles on one side of N add up to less than two right angles, then the lines L and M intersect on that side of N.

The fourth and fifth axioms are illustrated in Figure 26. The fourth means that you can slide any right angle until it lies exactly on any other. As for the fifth, because the angles marked a and β add up to less than 180 degrees, it tells us that the lines L and M meet somewhere to the right of N. The fifth axiom is equivalent to the so-called 'parallel postulate', which asserts that, given any line L and any point x not lying on L, there is exactly one line M that goes through x and never meets L.

Euclid used these five axioms to build up the whole of geometry as it was then understood. Here, for example, is an outline of how to prove the well-known result that the angles of a triangle add up to 180 degrees. The first step is to show that if a line N meets two parallel lines L and M, then opposite angles are equal. That is, in something like Figure 27 one must have $a = a'$ and $\beta = \beta'$. This is a consequence of the fifth axiom. First, it tells us that $a' + \beta$ is at least 180, or otherwise L and M would have to meet (somewhere to the left of line N in the diagram). Since a and β together make a straight line, $\beta = 180 - a$, so it follows that $a' + (180 - a)$ is at least 180, which means that a' is at least as big as a. By the same argument, $a + \beta' = a + (180 - a')$ must be at least 180, so a is at least as big as a'. The only way this can happen is if a and a' are equal. Since $\beta = 180 - a$ and $\beta' = 180 - a'$, it follows that $\beta = \beta'$ as well.

Now let ABC be a triangle, and let the angles at A, B, and C be a, β, and γ respectively. By the second axiom we can extend the line segment AC to a line L. The parallel postulate tells us that there is a line M through B that does not meet L. Let a' and γ' be the angles marked in Figure 28. By what we have just proved, $a' = a$ and $\gamma' = \gamma$. It is clear that $a' + \beta + \gamma = 180$, since the three angles a', β, and γ' together make a straight line. Therefore $a + \beta + \gamma = 180$, as required.

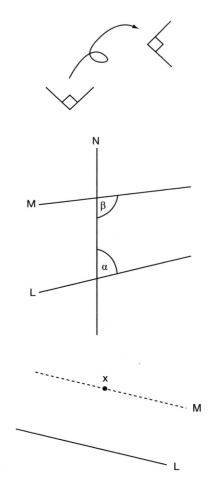

26. Euclid's fourth axiom and two versions of the fifth

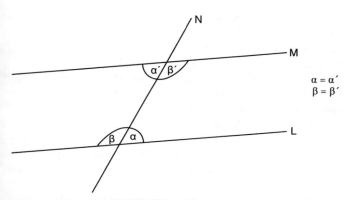

$\alpha = \alpha'$
$\beta = \beta'$

27. A consequence of Euclid's fifth axiom

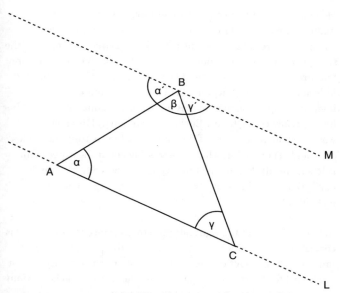

28. A proof that the angles of a triangle add up to 180 degrees

What does this argument tell us about everyday life? An obvious conclusion seems to be that if you take three points *A*, *B*, and *C* in space, and carefully measure the three angles of the triangle *ABC*, then they will add up to 180 degrees. A simple experiment will confirm this: just draw a triangle on a piece of paper, cut it out as neatly as you can, tear it into three pieces containing one corner each, place the corners together, and observe that the angles do indeed form a straight line.

If you are now convinced that there could not conceivably be a physical triangle with angles that fail to add up to 180 degrees, then you are in good company, as this was the conclusion drawn by everybody from Euclid in 300 BC to Immanuel Kant at the end of the 18th century. Indeed, so convinced was Kant that he devoted a significant part of his *Critique of Pure Reason* to the question of how one could be absolutely certain that Euclidean geometry was true.

However, Kant was mistaken: about thirty years later the great mathematician Carl Friedrich Gauss *could* conceive of such a triangle, as a result of which he actually measured the angles of the triangle formed by the mountain peaks of Hohenhagen, Inselberg, and Brocken in the kingdom of Hanover, to test whether they did indeed add up to 180 degrees. (This story is a famous one, but honesty compels me to note that there is some doubt about whether he was really trying to test Euclidean geometry.) His experiment was inconclusive because of the difficulty of measuring the angles accurately enough, but what is interesting about the experiment is less its result than the fact that Gauss bothered to attempt it at all. What could possibly be wrong with the argument I have just given?

Actually, this is not the right question to ask, since the *argument* is correct. However, since it rests on Euclid's five axioms, it does not imply anything about everyday life unless those axioms are true in everyday life. Therefore, by questioning the truth of Euclid's axioms one can question the *premises* of the argument.

But which of the axioms looks in the least bit dubious? It is difficult to find fault with any of them. If you want to join two points in the real world by a line segment, then all you have to do is hold a taut piece of string in such a way that it goes through both points. If you want to extend that line segment to a straight line, then you could use a laser beam instead. Similarly, there seems to be no difficulty with producing circles of any desired radius and centre, and experience shows that if you take two right-angled corners of paper, you can place one of them exactly over the other one. Finally, what is to stop two lines going off into the distance for ever, like a pair of infinitely long railway tracks?

The parallel postulate

Historically, the axiom that caused the most suspicion, or at least uneasiness, was the parallel postulate. It is more complicated than the other axioms, and involves the infinite in a fundamental way. Is it not curious, when one proves that the angles of a triangle add up to 180 degrees, that the proof should depend on what happens in the outermost reaches of space?

Let us examine the parallel postulate more carefully, and try to understand why it feels so obviously true. Perhaps in the back of our minds is one of the following arguments.

(1) Given a straight line L and a point x not on it, all you have to do to produce a parallel line through x is choose the line through x that goes *in the same direction as L*.

(2) Let y be another point, on the same side of L as x and at the same distance from L. Join x to y by a line segment (axiom 1) and then extend this line segment to a full line M (axiom 2). Then M will not meet L.

(3) Let M be the straight line consisting of all points on the same side of L as x and at the same distance. Obviously this does not meet L.

The arguments so far concern the *existence* of a line parallel to *L*. Here is a more complicated argument aimed at showing that there can be at most one such line, which is the other part of the parallel postulate.

(4) Join *L* to *M* by evenly spaced perpendicular line segments (making the railway tracks of Figure 29) with one of these segments through *x*. Now suppose that *N* is another line through *x*. On one side of *x*, the line *N* must lie between *L* and *M*, so it meets the next line segment at a point *u*, which is between *L* and *M*. Suppose for the sake of example that *u* is 1% of the way along the line segment from *M* to *L*. Then *N* will meet the next line segment 2% of the way along, and so on. Thus, after 100 segments, *N* will have met *L*. Since all we have assumed about *N* is that it is not *M*, it follows that *M* is the only line through *x* that does not meet *L*.

Finally, here is an argument that appears to show both the existence and the uniqueness of a line parallel to *L* through a given point.

(5) A point in the plane can be described by Cartesian coordinates. A (non-vertical) line *L* has an equation of the form $y = mx + c$. By

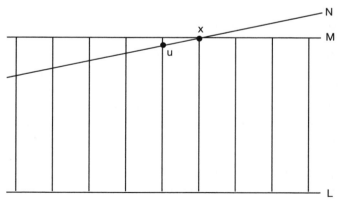

29. The uniqueness of parallel lines

varying c we can move L up and down. Obviously, no two of the lines thus produced can cross each other, and every point is contained in exactly one of them.

Notice that what I have just done is try to *prove* the parallel postulate, and that is exactly what many mathematicians before the 19th century tried to do. What they most wanted was to deduce it from the other four axioms, thus showing that one could dispense with it. However, nobody managed to do so. The trouble with the arguments I have just given, and others like them, is that they contain hidden assumptions which, when one makes them explicit, are not obvious consequences of Euclid's first four axioms. Though plausible, they are no *more* plausible than the parallel postulate itself.

Spherical geometry

A good way to bring these hidden assumptions out into the open is to examine the same arguments applied in a different context, one in which the parallel postulate is definitely not true. With this in mind, let us think for a moment about the surface of a sphere.

It is not immediately obvious what it means to say that the parallel postulate is untrue on the surface of a sphere, because the surface of a sphere contains no straight lines at all. We shall get round this difficulty by applying an idea of fundamental importance in mathematics. The idea, which is a profound example of the abstract method at work, is to *reinterpret* what is meant by a straight line, so that the surface of a sphere does contain straight lines after all.

There is in fact a natural definition: a line segment from x to y is the shortest path from x to y *that lies entirely within the surface of the sphere*. One can imagine x and y as cities and the line segment as the shortest route that an aeroplane could take. Such a path will be part of a 'great circle', which means a circle obtained by taking a plane through the centre of the sphere and seeing where it cuts the surface (see Figure 30). An example of a great circle is the equator of the earth (which, for the purposes of discussion, I shall take to be a perfect sphere). Given the way that we have defined line segments, a great circle makes a good definition of a 'straight line'.

If we adopt this definition, then the parallel postulate is certainly false. For example, let L be the earth's equator and let x be a point in the northern hemisphere. It is not hard to see that any great circle through x will lie half in the northern hemisphere and half in the southern, crossing the equator at two points that are exactly opposite each other (see Figure 31). In other words, there is no line (by which I still mean great circle) through x that does not meet L.

This may seem a cheap trick: if I define 'straight line' in a new way, then it is not particularly surprising if the parallel postulate ceases to hold. But it is not meant to be surprising – indeed the definition was designed for that very purpose. It becomes interesting when we examine some of the attempted *proofs* of the parallel postulate. In each case, we will discover an assumption that is not valid for spherical geometry.

For example, argument (1) assumes that it is obvious what is meant by the phrase 'the same direction'. But on the surface of a sphere this is not obvious at all. To see this, consider the three points P, Q, and N shown in Figure 32. N is the North Pole, P lies on the equator, and Q also lies on the equator, a quarter of the way round from P. Also on Figure 32 is a small arrow at P, pointing along the equator towards Q. Now what arrow could one draw at Q going in the same direction? The natural direction to choose is still along the equator, away from P. What about an arrow at N in the same direction again? We could choose this as follows. Draw the line segment from P to N. Since the arrow at P is at right angles to this line segment, the arrow at N should be as well, which means, in fact, that it points down towards Q. However, we now have a problem, which is that

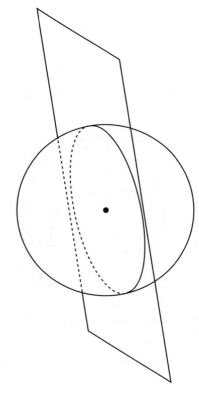

30. A great circle

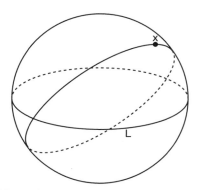

31. The parallel postulate is false for spherical geometry

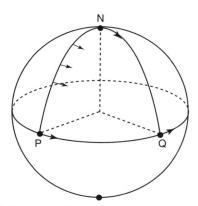

32. The phrase 'in the same direction as' does not make sense on the surface of a sphere

the arrow we have drawn at N does not point in the same direction as the one at Q.

The trouble with argument (2) is that it is not detailed enough. *Why* will the line M defined there not meet L? After all, if L and M are spherical lines then they *will* meet. As for argument (3), it assumes that M is a straight line. This is untrue for the sphere: if L is the equator and M consists of all points that are 1,000 miles north of the equator, then M is *not* a great circle. In fact, it is a line of constant latitude, which, as any pilot or sailor will tell you, does not give the shortest route between two points.

Argument (4) is a little different, since it is concerned with the uniqueness of parallel lines rather than their existence. I shall discuss this in the next section. Argument (5) makes a huge assumption: that space can be described by Cartesian coordinates. Again, this is untrue of the surface of the sphere.

The point about introducing the sphere is that it enables us to isolate from each of the arguments (1), (2), (3) and (5) some assumption that effectively says, 'The geometry we are doing is not spherical geometry.' You might wonder what is wrong with this: after all, we are *not* doing spherical geometry. You might also wonder how one could ever hope to show that the parallel postulate does not follow from the rest of Euclid's axioms, if indeed it doesn't. It is no good saying that mathematicians have tried to deduce it for centuries without success. How can we be sure that some young genius in two hundred years' time will not have a wonderful new idea which finally leads to a proof?

This question has a beautiful answer, at least in principle. Euclid's first four axioms were devised to describe the geometry of an infinite, flat, two-dimensional space, but we are not obliged to interpret them that way, unless, of course, this flatness actually follows from the axioms. If we could somehow reinterpret (one might almost say 'misinterpret') the axioms by giving new meanings to phrases such as 'line segment', rather as we have done with spherical geometry, and if when we had done so we found that the first four axioms were true but the parallel postulate was false, then we would have shown that the parallel postulate does not follow from the other axioms.

To see why this is so, imagine a purported proof, which starts with Euclid's first four axioms and arrives, after a series of strict logical steps, at the parallel postulate. Since the steps follow as a matter of logic, they will remain valid even when we have given them their new interpretation. Since the first four axioms are true under the new interpretation and the parallel postulate is not true, there must be a mistake in the argument.

Why do we not just use spherical geometry as our reinterpretation? The reason is that, unfortunately, not all of the first four of Euclid's axioms are true in the sphere. For example, a sphere does not contain circles of arbitrarily large radius, so axiom 3 fails, and there is not just one shortest route from the North to the South Pole, so axiom 1 is not true either. Hence, although spherical geometry helped us understand the defects of certain attempted proofs of the parallel postulate, it still leaves open the possibility that some other proof might work. Therefore, I shall turn to another reinterpretation, called hyperbolic geometry. The parallel postulate will again be false, but this time axioms 1 to 4 will all be true.

Hyperbolic geometry

There are several equivalent ways of describing hyperbolic geometry; the one I have chosen is known as the disc model, which was discovered by the great French mathematician Henri Poincaré. While I cannot define it precisely in a book like this, I can at least explain some of its main features and discuss what it tells us about the parallel postulate.

Understanding the disc model is more complicated than understanding spherical geometry because one has to reinterpret not only the terms 'line' and 'line segment', but also the idea of distance. On the surface of the sphere, distance has an easily grasped definition: the distance between two points x and y is the shortest possible length of a path from x to y that lies within the surface of the sphere. Although a similar definition holds for hyperbolic geometry, it is not obvious, for reasons that will become clear, what the shortest path is, or indeed what the length of *any* path is.

Figure 33 shows a tessellation of the hyperbolic disc by regular pentagons. Of course, this statement has to be explained, since it is untrue if we understand distance in the usual way: the edges of these 'pentagons' are visibly not straight and do not have the same length. However, distances in the hyperbolic disc are not defined in the usual way, and become larger, relative to normal distance, as you approach the boundary. Indeed, they become so much larger that the boundary is, despite appearances, infinitely far from the centre. Thus, the reason that the pentagon marked with an asterisk appears to have one side longer than all the others is that that side is closer to the centre. The other sides may look shorter, but hyperbolic distance is defined in such a way that this apparent shortness is exactly compensated for by their being closer to the edge.

33. **A tessellation of the hyperbolic plane by regular pentagons**

If this seems confusing and paradoxical, then think of a typical map of the world. As everyone knows, because the world is round and the map is flat, distances are necessarily distorted. There are various ways of carrying out the distortion, and in the most common one, Mercator's projection, countries near the poles appear to be much larger than they really are. Greenland, for example, seems to be comparable in size to the whole of South America. The nearer you are to the top or bottom of such a map, the smaller the distances are, compared with what they appear to be.

A well-known effect of this distortion is that the shortest route between two points on the earth's surface appears, on the map, to be curved. This phenomenon can be understood in two ways. The first is to forget the map and visualize a globe instead, and notice that if you have two points in the northern hemisphere, the first a long way east of the other (a good example is Paris and Vancouver), then the shortest route from the first to the second will pass close to the North Pole rather than going due west. The second is to argue from the original map and reason that if distances near the top of the map are shorter than they appear, then one can shorten the journey by going somewhat north as well as west. It is difficult to see in this way precisely what the shortest path will be, but at least the principle is clear that a 'straight line' (from the point of view of spherical distances) will be curved (from the point of view of the distances on the actual map).

As I have said, when you approach the edge of the hyperbolic disc, distances become *larger* compared with how they look. As a result of this, the shortest path between two points has a tendency to deviate towards the centre of the disc. This means that it is not a straight line in the usual sense (unless that line happens to pass exactly through the centre). It turns out that a hyperbolic straight line, that is, a shortest path from the point of view of hyperbolic geometry, is the arc of a circle that meets the boundary of the main circle at right angles (see Figure 34). If you now look again at the pentagonal tessellation of Figure 35, you will see that the edges of

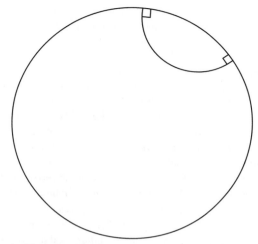

34. A typical hyperbolic line

the pentagons, though they do not appear straight, are in fact hyperbolic line segments since they can be extended to hyperbolic straight lines, according to the definition I have just given. Similarly, although the pentagons do not seem to be all of the same size and shape, they are, since the ones near the edge are far bigger than they seem – the opposite of what happens with Greenland. Thus, just like Mercator's projection, the disc model is a distorting 'map' of actual hyperbolic geometry.

It is natural to ask at this point what actual hyperbolic geometry is like. That is, what is the distorting map a map *of*? What stands in relation to the disc model as the sphere does to Mercator's projection? The answer to this is rather subtle. In a way it is a fluke that spherical geometry can be realized as a surface that sits in three-dimensional space. If we had *started* with Mercator's projection, with its strange notion of distances, without knowing that what we had was a map of the sphere, then we would have been surprised and delighted to discover that there happened to be a beautifully symmetrical surface in space, a map of this map, so to speak, where distances were particularly simple, being nothing but the lengths of shortest paths in the usual, easily understood sense.

Unfortunately, nothing like this exists for hyperbolic geometry. Yet, curiously, this does not make hyperbolic geometry any less real than spherical geometry. It makes it harder to understand, at least initially, but as I stressed in Chapter 2 the reality of a mathematical concept has more to do with what it does than with what it is. Since it is possible to say what the hyperbolic disc does (for example, if you asked me what it would mean to rotate the pentagonal tessellation through 30 degrees about one of the vertices of the central pentagon, then I could tell you), hyperbolic geometry is as real as any other mathematical concept. Spherical geometry may be easier to understand from the point of view of three-dimensional Euclidean geometry, but this is not a fundamental difference.

Another of the properties of hyperbolic geometry is that it satisfies the first four of Euclid's axioms. For example, any two points can be joined by exactly one hyperbolic straight line segment (that is, arc of a circle that cuts the main circle at right angles). It may seem as though you cannot find a circle of large radius about any given point, but to think that is to have forgotten that distances become larger near the edge of the disc. In fact, if a hyperbolic circle almost brushes the edge, then its radius (its hyperbolic radius, that is) will be very large indeed. (Hyperbolic circles happen to look like ordinary circles, but their centres are not where one expects them to be. See Figure 35.)

As for the parallel postulate, it is false for hyperbolic geometry, just as we hoped. This can be seen in Figure 36, where I have marked three of the (hyperbolic) lines L, M_1, and M_2. The lines M_1 and M_2 meet at a point marked x, but neither of them meets L. Thus, there are two lines through x (and in fact infinitely many) that do not meet L. This contradicts the parallel postulate, which stipulates that there should be only one. In other words, in hyperbolic geometry we

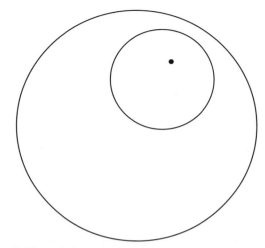

35. A typical hyperbolic circle, and its centre

have exactly the alternative interpretation of the Euclidean axioms that we were looking for in order to show that the parallel postulate was not a consequence of the other four axioms.

Of course, I have not actually proved in this book that hyperbolic geometry has all the properties I have claimed for it. To do so takes a few lectures in a typical university mathematics course, but I can at least say more precisely how to define hyperbolic distance. To do this, I must specify *by how much* distances near the edge of the disc are larger than they appear. The answer is that hyperbolic distances at a point P are larger than 'normal' distances by $1/d^2$, where d is the distance from P to the boundary of the circle. To put that another way, if you were to move about in the hyperbolic disc, then your speed as you passed P would, according to hyperbolic notions of distance, be $1/d^2$ times your apparent speed, which means that if you maintained a constant hyperbolic speed, you would appear to move more and more slowly as you approached the boundary of the disc.

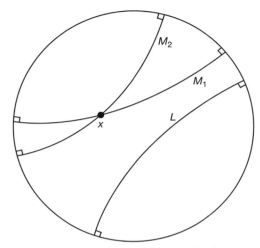

36. The parallel postulate is false in the hyperbolic plane

Just before we leave hyperbolic geometry, let us see why argument (4) that I gave earlier fails to prove the uniqueness of parallel lines. The idea was that, given a line L, a point x not on L, and a line M through x that did not meet L, one could join L to M by several line segments that were perpendicular to both L and M, dividing up the space between L and M into rectangles. It seems obvious that one can do this, but in the hyperbolic world it is not possible because the angles of a quadrilateral always add up to *less* than 360 degrees. In other words, in the hyperbolic disc the rectangles needed for the argument simply do not exist.

How can space be curved?

One of the most paradoxical-sounding phrases in mathematics (and physics) is 'curved space'. We all know what it means for a line or surface to be curved, but space itself just *is*. Even if one could somehow make sense of the idea of three-dimensional curviness, the analogy with a curved surface suggests that we would not be able to see for ourselves whether space was curved unless we could step out into a fourth dimension to do so. Perhaps we would then discover that the universe was the three-dimensional surface of a four-dimensional sphere (a notion I explained in Chapter 5), which at least *sounds* curved.

Of course, all this is impossible. Since we do not know how to stand outside the universe – the very idea is almost a contradiction in terms – the only evidence we can use comes from within it. So what evidence might persuade us that space is curved?

Yet again, the question becomes easier if one takes an abstract approach. Instead of engaging in extraordinary mental gymnastics as we try to apprehend the true nature of curvy space, let us simply follow the usual procedure for generalizing mathematical concepts. We understand the word 'curved' when it applies to two-dimensional surfaces. In order to use it in an unfamiliar context, that is, a three-dimensional one, we must try to find *properties* of curved surfaces that will generalize easily, just as we did in order to define $2^{3/2}$, or five-dimensional cubes, or the dimension of the Koch snowflake. Since the sort of property we wish to end up with is one that can be detected from *within* space, we ought to look at ways of detecting the curvature of a surface that do not depend on standing outside it.

How, for example, can we convince ourselves that the surface of the earth is curved? One way is to go up in a space shuttle, look back, and see that it is approximately spherical. However, the following experiment, which is much more two-dimensional, would also be very persuasive. Start at the North Pole and travel due south for about 6,200 miles, having marked your initial direction. Then turn to your left and go the same distance again. Then turn to your left and go the same distance one more time. 6,200 miles is roughly the distance from the North Pole to the equator, so your journey will have taken you from the North Pole to the equator, a quarter of the way round the equator, and back to the North Pole again. Moreover, the direction at which you arrive back will be at right angles to your starting direction. It follows that on the earth's surface there is an equilateral triangle with all its angles equal to a right angle. On a flat surface, the angles of an equilateral triangle have to be 60 degrees, as they are all equal and add up to 180, so the surface of the earth is not flat.

Thus, one way of demonstrating that a two-dimensional surface is curved, from within that surface, is to find a triangle whose angles do not add up to 180 degrees, and this is something that can be attempted in three dimensions as well. I have concentrated in this chapter on Euclidean, spherical, and hyperbolic geometry in two dimensions, but they can be generalized quite easily to three dimensions. If we measure the angles of triangles in space and find that they add up to more than 180 degrees, then that will suggest that space is more like a three-dimensional version of the surface of a sphere than like the sort of space that can be described by three Cartesian coordinates.

If this happens, then it seems reasonable to say that space is positively curved. Another feature that one would expect of such a space is that lines that started off in the same direction would converge and eventually meet. Still another is that the circumference of a circle of radius r would not be $2\pi r$, but a little less.

You may be tempted to point out that space as we know it does not have these peculiarities. Lines that begin in the same direction continue in the same direction, and the angles of triangles and circumferences of circles are what they ought to be. In other words, it appears that, even though it is logically possible for space to be curved, as a matter of fact it is flat. However, it could be that space appears to be flat to us only because we inhabit such a small part of it, just as the earth's surface appears to be flat, or rather flat with bumps of various sizes, to somebody who has not travelled far.

In other words, it may be that space is only *roughly* flat. Perhaps if we could form a very large triangle then we would find that its angles did not add up to 180 degrees. This, of course, was what Gauss attempted, but his triangle was nowhere near large enough. However, in 1919, one of the most famous scientific experiments of all time showed that the idea of curved space was not just a fantasy of mathematicians, but a fact of life. According to Einstein's general theory of relativity, which was published four years earlier, space is curved by gravity, and therefore light does not always travel in a straight line, at least as Euclid would understand the term. The effect is too small to be detected easily, but the opportunity came in 1919 with a total eclipse of the sun, visible from Principe Island in the Gulf of Guinea. While it was happening, the physicist Arthur Eddington took a photograph that showed the stars just next to the sun in not quite their expected places, exactly as Einstein's theory had predicted.

Though it is now accepted that space (or, more accurately, spacetime) is curved, it could be that, like the mountains and valleys on the earth's surface, the curvature that we observe is just a small perturbation of a much larger and more symmetrical shape. One of the great open questions of astronomy is to determine the *large-scale* shape of the universe, the shape that it would have if one ironed out the curves due to stars, black holes, and so on. Would it still be curved, like a large sphere, or would it be flat, as one more naturally, but quite possibly wrongly, imagines it?

A third possibility is that the universe is *negatively* curved. This means, not surprisingly, more or less the opposite of positively curved. Thus, evidence for negative curvature would be that the angles of a triangle added up to *less* than 180 degrees, that lines starting in the same direction tended to diverge, or that the circumference of a circle of radius r was *larger* than $2\pi r$. This sort of behaviour occurs in the hyperbolic disc. For example, Figure 37 shows a triangle whose angles add up to significantly less than 180 degrees. It is not hard to generalize the sphere and the hyperbolic

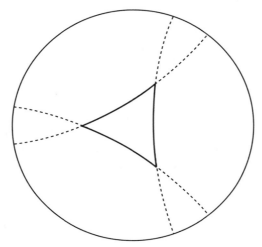

37. A hyperbolic triangle

disc to higher-dimensional analogues, and it could be that
hyperbolic geometry is a better model for the large-scale shape of
spacetime than either spherical or Euclidean geometry.

Manifolds

A closed surface means a two-dimensional shape that has no
boundary. The surface of a sphere is a good example, as is a torus
(the mathematical name for the shape of the surface of a quoit, or a
ring-shaped doughnut). As the discussion of curvature made clear,
it can be useful to think about such surfaces without reference to
some three-dimensional space in which they live, and this becomes
even more important if we want to generalize the notion of a closed
surface to higher dimensions.

It is not just mathematicians who like to think about surfaces in a
purely two-dimensional way. For example, the geometry of the
United States is significantly affected by the curvature of the earth,

but if one wishes to design a useful road map, it does not have to be printed on a single large curved piece of paper. Much more practical is to produce a book with several pages, each dealing with a small part of the country. It is best if these parts overlap, so that if a town lies inconveniently near the edge of one page, there will be another page where it doesn't. Moreover, at the edges of each page will be an indication of which other pages represent overlapping regions and how the overlap works. Because of the curvature of the earth, none of the pages will be exactly accurate, but one can include lines of constant latitude and longitude to indicate the small distortion, and in that way the geometry of the United States can be encapsulated in a book of flat pages.

There is nothing in principle to stop one producing an atlas that covers the whole world in similar detail (though many pages would be almost entirely blue). Therefore, the mathematical properties of a sphere can in a way be encapsulated in an atlas. If you want to answer geometrical questions about the sphere, are completely unable to visualize it, but have an atlas handy, then, with a bit of effort, you will be able to do it. Figure 38 shows a nine-page atlas, not of a sphere but of a torus. To see how it corresponds to a doughnut shape, imagine sticking the pages together to make one large page, then joining the top and bottom of the large page to form a cylinder, and finally bringing the two ends of the cylinder together and joining them.

One of the most important branches of mathematics is the study of objects known as manifolds, which result from generalizing these ideas to three or more dimensions. Roughly speaking, a d-dimensional manifold is any geometrical object in which every point is surrounded by a region that closely resembles a small piece of d-dimensional space. Since manifolds become much harder to visualize as the number of dimensions increases, the idea of an atlas becomes correspondingly more useful.

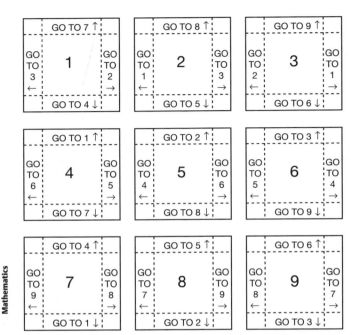

38. An atlas of a torus

Let us think for a moment what an atlas of a three-dimensional manifold would be like. The pages would of course have to be three-dimensional, and like the pages of a road map they would be flat. By that I mean that they would be chunks of familiar Euclidean space; one could require them to be cuboids, but this is not very important mathematically. Each of these three-dimensional 'pages' would be a map of a small part of the manifold, and one would carefully specify how the pages overlapped. A typical specification might be something like that the point (x, y, z) towards a particular edge of page A corresponded to the point $(2y, x, z - 4)$ on page B.

Given such an atlas, how could one imagine moving about in the manifold? The obvious way would be to think of a point moving in one of the pages. If this point ever reached the edge of the page, there would be another page on which the same part of the manifold was represented, but where the point was not right at the edge, so one could turn to that page instead. Thus, the entire geometry of a manifold can be formulated in terms of an atlas, so that it is not necessary to think of the manifold as 'really' being a three-dimensional surface lying in a four-dimensional space. In fact, some three-dimensional manifolds cannot even be made to fit into four dimensions.

This idea of an atlas raises some natural questions. For example, although it enables us to say what happens if we move around in the manifold, how do we get from that information, which may be contained in a large number of pages with very complicated rules for how they overlap, to some feeling for the basic 'shape' of the manifold? How can we tell when two different atlases are actually atlases of the same manifold? In particular, is there some easy way of telling, by looking at a three-dimensional atlas, whether the manifold it represents is the three-dimensional surface of a four-dimensional sphere? A precise formulation of this last question, known as the Poincaré conjecture, is an open problem, for the solution of which a reward of one million dollars has been offered (by the Clay Mathematics Institute).

Chapter 7
Estimates and approximations

Most people think of mathematics as a very clean, exact subject. One learns at school to expect that if a mathematical problem can be stated succinctly, then it will probably have a short answer, often given by a simple formula. Those who continue with mathematics at university level, and particularly those who do research in the subject, soon discover that nothing could be further from the truth. For many problems it would be miraculous and totally unexpected if somebody were to find a precise formula for the solution; most of the time one must settle for a rough estimate instead. Until one is used to estimates, they seem ugly and unsatisfying. However, it is worth acquiring a taste for them, because not to do so is to miss out on many of the greatest theorems and most interesting unsolved problems in mathematics.

A simple sequence not given by a simple formula

Let $a_1, a_2, a_3, \ldots$ be real numbers generated by the following rule. The first number, a_1, equals 1, and thereafter each number equals the previous number plus its square root. In other words, for every n we let $a_{n+1} = a_n + \sqrt{a_n}$. This simply stated rule raises an obvious question: what is the number a_n?

To get a feel for the question, let us work out a_n for a few small values of n. We have $a_2 = 1 + \sqrt{1} = 1 + 1 = 2$. Then

$$a_3 = a_2 + \sqrt{a_2} = 2 + \sqrt{2}$$
$$a_4 = a_3 + \sqrt{a_3} = 2 + \sqrt{2} + \sqrt{2 + \sqrt{2}}$$
$$a_5 = a_4 + \sqrt{a_4} = 2 + \sqrt{2} + \sqrt{2 + \sqrt{2}} + \sqrt{2 + \sqrt{2} + \sqrt{2 + \sqrt{2}}}$$

and so on. Notice that the expressions on the right-hand side do not seem to simplify, and that each new one is twice as long as the previous one. From this observation it follows quite easily that the expression for a_{12} would involve 1,024 occurrences of the number 2, most of them deep inside a jungle of square-root signs. Such an expression would not give us much insight into the number a_{12}.

Should we therefore abandon any attempt to understand the sequence? No, because although there does not appear to be a good way of thinking about the *exact* value of a_n, except when n is very small, that does not rule out the possibility of obtaining a good estimate. Indeed, a good estimate may in the end be more useful. Above, I have written an exactly correct expression for a_5, but does that make you understand a_5 better than the information that a_5 is about seven and a half?

Let us therefore stop asking what a_n is and instead ask roughly how big a_n is. That is, let us try to find a simple formula that gives a good approximation to a_n. It turns out that such a formula exists: a_n is roughly $n^2/4$. It is a little tricky to prove this rigorously, but to see why it is plausible, notice that

$$(n + 1)^2/4 = (n^2 + 2n + 1)/4 = n^2/4 + n/2 + 1/4 = n^2/4 + \sqrt{n^2/4} + 1/4.$$

That is, if $b_n = n^2/4$, then $b_{n+1} = b_n + \sqrt{b_n} + 1/4$. If it were not for the '+ 1/4', this would tell us that the numbers b_n were generated exactly as the a_n are. However, when n is large, the addition of 1/4 is 'just a small perturbation' (this is the part of the proof that I am leaving out), so the b_n are generated *approximately* correctly, from which it can be deduced that b_n, that is, $n^2/4$, gives a good approximation to a_n, as I claimed.

Ways of approximating

It is important to specify what counts as a good approximation when making an assertion like this, because standards vary from context to context. If one is trying to approximate a sequence of steadily increasing numbers like $a_1, a_2, a_3, \ldots$ by a more easily defined sequence $b_1, b_2, b_3, \ldots$, then the best sort of approximation one can hope for, which is very rarely achieved, is one for which the difference between a_n and b_n is always below some fixed number – such as 1,000, for example. Then, as a_n and b_n become large, their *ratio* becomes very close to 1. Suppose, for example, that at some point $a_n = 2408597348632498758828$ and $b_n = 2408597348632498759734$. Then $b_n - a_n = 906$, which, though quite a large number, is tiny by comparison with a_n and b_n. If b_n approximates a_n in this sense, one says that a_n and b_n are 'equal up to an additive constant'.

Another good sort of approximation is one for which the ratio of a_n and b_n becomes very close to 1 as n gets large. This is true when a_n and b_n are equal up to an additive constant, but it is also true in other circumstances. For example, if $a_n = n^2$ and $b_n = n^2 + 3n$ then the ratio b_n/a_n is $1 + 3/n$, which is close to 1 for large n, even though the *difference* between a_n and b_n is $3n$, which is large.

Often even this is far too much to hope for, and one is happy to settle for still weaker notions of approximation. A common one is to regard a_n and b_n as approximately equal if they are 'equal up to a *multiplicative* constant'. This means that neither a_n/b_n nor b_n/a_n ever exceeds some fixed number – again, something like 1,000 would be a possibility, though the smaller the better. In other words, now it is not the difference between a_n and b_n that is kept within certain limits, but their ratio.

It may seem perverse to regard one number as being roughly the same as another number that is 1,000 times larger. But that is because we are used to dealing with small numbers. Of course, nobody would regard 17 as being roughly the same as 13,895, but it is not quite as ridiculous to say that two numbers like

290475629408976156238954534598760889079687234751434875775468

and

360982345987209761234987098249863408762345779678458734598716464

are, broadly speaking, of the same *sort* of size. Though the second is over 1,000 times larger, they both have about the same number of digits – between 60 and 65. In the absence of other interesting properties, that may well be all we care about.

When even *this* degree of approximation is asking too much, it is often still worthwhile to try to find two sequences b_1, b_2, b_3, . . . and c_1, c_2, c_3, . . . for which one can prove that b_n is always less than a_n, and c_n always greater. Then one says that b_n is a 'lower bound' for a_n, and c_n is an 'upper bound'. For example, a mathematician trying to estimate some quantity a_n might say, 'I don't know, even approximately, what a_n is, but I can prove that it is at least $n^2/2$ and no bigger than n^3.' If the problem is difficult enough, a theorem like this can be a significant achievement.

All you need to know about logarithms, square roots etc.

Part of the reason that the estimates and approximations that pervade mathematics are not well known outside the discipline is that in order to talk about them one uses phrases like 'about as fast as log n' or 'the square root of t, to within a constant', which mean little to most people. Fortunately, if one is concerned only with the *approximate* values of logarithms or square roots of large numbers, then they can be understood very easily, and so can this sort of language.

If you have two large positive integers m and n and you want a rough and ready estimate of their product mn, then what should you do? A good start is to count the digits of m and the digits of n. If m has h digits and n has k digits, then m lies between 10^{h-1} and 10^h, and n lies between 10^{k-1} and 10^k, which means that mn lies between 10^{h+k-2} and 10^{h+k}. Thus, merely by counting the digits of m and n, you can determine mn 'to within a factor of 100' – meaning that mn must lie between two numbers, 10^{h+k-2} and 10^{h+k}, and 10^{h+k} is only 100 times bigger than 10^{h+k-2}. If you compromise and go for 10^{h+k-1} as your estimate, then it will differ from mn by a factor of at most 10.

In other words, if you are interested in numbers only 'up to a multiplicative constant', then multiplication suddenly becomes very easy: take m and n, count their combined digits, subtract one (if you can be bothered), and write down a number with that many digits. For example, 1293875 (7 digits) times 20986759777 (11 digits) is in the region of 10000000000000000 (17 digits). If you want to be a little more careful, you can note that the first number begins with a 1 and the second with a 2, which means that 20000000000000000 is a better estimate, but for many purposes such precision is unnecessary.

Since approximate multiplication is easy, so is approximate squaring – just replace your number by a new one with twice as many digits. From this it follows that *halving* the number of digits of n approximates the square root of n. Similarly, dividing the number of digits by 3 approximates the cube root. More generally, if n is a large integer and t is any positive number, then n^t will have about t times as many digits as n.

What about logarithms? From the approximate point of view they are very simple indeed: the logarithm of a number is roughly the number of digits it has. For example, the logarithms of 34587 and 492348797548735 are about 5 and 15 respectively.

Actually, counting the digits of a number approximates its base-10 logarithm, the number you get by pressing LOG on a pocket calculator. Normally, when mathematicians talk about logarithms, they mean so-called 'natural' logarithms, which are logarithms to base e. Though the number e is indeed very natural and important, all we need to know here is that the natural logarithm of a number, the number you get by pressing LN on a calculator, is roughly the number of digits it has, multiplied by about 2.3. Thus, the natural logarithm of 2305799985748 is about $13 \times 2.3 = 29.9$. (If you know about logarithms, you will see that what you should really multiply by is $\log_e 10$.)

This process can be reversed. Suppose you have a number t and you know that it is the natural logarithm of another number n. This number n is called the *exponential* of t and is written e^t. What will n be, roughly? Well, to get t from n we count the number of digits of n and multiply by 2.3. Hence, the number of digits of n must be about $t/2.3$. This determines n, at least approximately.

The main use of the approximate definitions I have just given is that they enable one to make comparisons. For example, it is now clear that the logarithm of a large number n will be much smaller than its cube root: if n has 75 digits, for example, then its cube root will be very large – it has about 25 digits – but its natural logarithm will be only about $75 \times 2.3 = 172.5$. Similarly, the exponential of a number m will be much bigger than a power such as m^{10}: for example, if m has 50 digits, then m^{10} has around 500 digits, but the number of digits of e^m is about $m/2.3$, which is far bigger than 500.

The following table shows the approximate results of applying various operations to the number $n = 941192$. I have not included e^n, because if I had then I would have been obliged to change the title of this book.

n	941192
n^2	885842380864
$\sqrt{n}$	970.15
$\sqrt[3]{n}$	98
$\log_e n$	13.755
$\log_{10} n$	5.974

The prime number theorem

A prime number is a whole number greater than 1 that is divisible by no other whole numbers, with the obvious exceptions of 1 and itself. The prime numbers less than 150 are 2, 3, 5, 7, 11, 13, 17, 19, 23, 29, 31, 37, 43, 47, 53, 59, 61, 67, 71, 73, 79, 83, 89, 97, 101, 103, 107, 109, 113, 127, 131, 137, 139, and 149. All other numbers less than 150 can be factorized: for example, $91 = 7 \times 13$. (You may be worried about the seemingly arbitrary exclusion of 1 from the definition of a prime. This does not express some deep fact about numbers: it just happens to be a useful convention, adopted so that there is only one way of factorizing any given number into primes.)

The primes have tantalized mathematicians since the Greeks, because they appear to be somewhat randomly distributed but not completely so. Nobody has found a simple rule that tells you what the nth largest prime is (of course one could laboriously write out a list of the first n primes, but this hardly counts as a simple rule, and would be completely impractical if n was large), and it is most unlikely that there is one. On the other hand, an examination of even the first 150 primes reveals some interesting features. If you work out the differences between successive primes, then you obtain the following new list: 1, 2, 2, 4, 2, 4, 2, 4, 6, 2, 6, 6, 4, 6, 6, 2, 6, 4, 2, 6, 4, 6, 8, 4, 2, 4, 2, 4, 14, 4, 6, 2, 10. (That is, $1 = 3 - 2$, $2 = 5 - 3$, $2 = 7 - 5$, $4 = 11 - 7$, and so on.) This list is still somewhat disorderly, but the numbers in it have a tendency, just about discernible, to get gradually larger. Obviously they do not increase steadily, but numbers such as 10 and 14 do not appear until quite late on, while the first few are all 4 or under.

If you were to write out the first thousand primes, then the tendency for the gaps between successive ones to get larger would become more obvious. In other words, large primes appear to be thinner on the ground than small ones. This is exactly what one would expect, because there are more ways for a large number to *fail* to be prime. For example, one might guess that the number 10,001 was prime, especially as it is not divisible by 2, 3, 5, 7, 11, 13, 17, or 19 – but in fact it equals 73×137.

No self-respecting mathematician will be content with the mere observation (not even properly proved) that large primes are rarer than small ones. He or she will want to know *how much* rarer they are. If you choose a number at random between 1,000,001 and 1,010,000, then what are the chances that it will be a prime? In other words, what is the 'density' of primes near 1,000,000? Is it fantastically small or only quite small?

The reason such questions rarely occur to people who have not been exposed to university-level mathematics is that they lack the language in which to formulate and think about them. However, if you have understood this chapter so far, you are in a position to appreciate one of the greatest achievements of mathematics: the prime number theorem. This states that the density of primes near a number n is around $1/\log_e n$ – that is, one divided by the natural logarithm of n.

Consider once again the chances that a random number between 1,000,001 and 1,010,000 is prime. The numbers in this interval are all roughly equal to a million. The prime number theorem says that the density will therefore be around 1 divided by the natural logarithm of a million. The logarithm to base 10 is 6 (in this case, counting the digits would give 7, but since we know the exact answer we may as well use it), so the natural logarithm is about 6×2.3, or 13.8. Therefore, about 1 in 14 numbers between 1,000,001 and 1,010,000 is a prime, which works out at a little over 7% of them. By contrast, the number of primes less than 100 is 24, or about a quarter of the total, which illustrates how the density drops as they get larger.

Given the sporadic, random-like quality of the primes, it is quite surprising how much can be proved about them. Interestingly, theorems about the primes are usually proved by *exploiting* this seeming randomness. For example, a famous theorem of Vinogradov, proved in 1937, states that every large enough odd number can be written as the sum of three prime numbers. I cannot explain in this book how he proved this, but what he did not do is find a *method* for expressing odd numbers as sums of three primes. Such an approach would be almost certain to fail, because of the difficulty of generating even the primes themselves. Instead, building on work of Hardy and Littlewood, he argued roughly as follows. If you were to choose a genuinely random sequence of numbers of about the same density as the primes, then some elementary probability theory shows that you would almost certainly be able to write all large enough numbers as the sum of three members of your sequence. In fact, you would be able to do it in many different ways. Because the primes are random-like (the hard part of the proof is to say what this means and then prove it rigorously), their behaviour is similar to that of the random sequence, so all large enough numbers are the sum of three primes, also in many different ways. Just to illustrate this phenomenon, here are all the ways of writing 35 as a sum of three primes:

$$35 = 2 + 2 + 31 = 3 + 3 + 29 = 3 + 13 + 19 = 5 + 7 + 23$$

$$= 5 + 11 + 19 = 5 + 13 + 17 = 7 + 11 + 17 = 11 + 11 + 13.$$

Much research on prime numbers has this sort of flavour. You first devise a probabilistic model for the primes – that is, you pretend to yourself that they have been selected according to some random procedure. Next, you work out what would be true if the primes really were generated randomly. That allows you to guess the answers to many questions. Finally, you try to show that the model is realistic enough for your guesses to be approximately correct. Notice that such an approach would be impossible if you were forced to give exact answers at every stage of the argument.

It is interesting that the probabilistic model is a model not of a physical phenomenon, but of another piece of mathematics. Although the prime numbers are rigidly determined, they somehow feel like experimental data. Once we regard them that way, it becomes tempting to devise simplified models that allow us to predict what the answers to certain probabilistic questions are likely to be. And such models have indeed sometimes led people to proofs valid for the primes themselves.

Although this style of argument has had some notable successes, it has left open many famous problems. For example, Goldbach's conjecture asserts that every even number greater than 4 is the sum of two odd primes. This conjecture appears to be much more difficult than the three-primes question answered by Vinogradov. There is also the twin-primes conjecture, which states that there are infinitely many pairs of primes separated by 2, such as 17 and 19, or 137 and 139. Another way of putting this is that if you write out the successive differences, as I did above, then the number 2 keeps appearing for ever (though more and more rarely).

Perhaps the most famous open problem in mathematics is the Riemann hypothesis. This has several equivalent formulations. One of them concerns the accuracy of the estimate given by the prime number theorem. As I have said, the prime number theorem tells you the approximate density of the primes near any given number. From this information one can calculate roughly how many prime numbers there are up to any given number n. But how rough is rough? If $p(n)$ is the true number of primes up to n and $q(n)$ is the estimate suggested by the prime number theorem, then the Riemann hypothesis asserts that the difference between $p(n)$ and $q(n)$ will be not much larger than $\sqrt{n}$. If that sort of accuracy could be proved to hold, then it would have many applications, but what is known to date is far weaker.

Sorting algorithms

Another area of mathematics that is full of rough estimates is theoretical computer science. If one is writing a computer program to perform a certain task, then it is a good idea to design it in such a way that it will run as quickly as possible. Theoretical computer scientists ask the question: what is the fastest one could possibly hope for?

It is almost always unrealistic to ask for an exact answer to this question, so one tries to prove statements like, 'The following algorithm runs in about n^2 steps when the input size is n.' From this one can conclude that a typical PC will be able to handle an input size (roughly speaking, how much information you want it to analyse) of 1,000 but not one of 1,000,000. Thus, such estimates have a practical importance.

One very useful task that computers can do is known as sorting – that is, putting a large number of objects in order according to a given criterion. To think about this, imagine that you wish to arrange a collection of objects (not necessarily inanimate – they might, for example, be candidates for a job) in order of preference. Suppose that you cannot assign a numerical value to the amount that you like any given object, but that, given any two objects, you can always decide which you prefer. Suppose also that your preferences are consistent, in the sense that you never prefer A to B, B to C, and C to A. If you do not wish to spend long on the task, then it makes sense to try to minimize the number of comparisons you make.

When the number of objects is very small, it is easy to work out how to do this. For example, if there are two objects, then you must make at least one comparison, and once you have made it you know what order they are in. If there are three objects A, B, and C, then one comparison will not be enough, but you must start with some comparison and it doesn't matter which. Suppose, for the sake of argument, that you compare A with B and prefer A. Now you must compare one of A and B with C. If you compare A with C and prefer C to A, then you know that your order of preference is C, A, B. However, if, as may happen, you find that you prefer A to C, then all you know about B and C is that you prefer A to either of them. Then a third comparison will be needed so that B and C can be put in order. Hence, three comparisons are always sufficient and sometimes necessary.

What happens with four objects A, B, C, and D? The analysis becomes more complicated. You may as well start by comparing A with B. But once you have done that, there are two genuinely different possibilities for the next comparison. Either you can compare one of A and B with C, or you can compare C with D, and it is not clear which is the better idea.

Suppose you compare B with C. If you are lucky, then you will now be able to put A, B, and C in order. Suppose that this order is A, B, C. It then remains to see where D fits in. The best thing to do first is to compare D with B. After that, all you have to do is compare D with A (if you preferred D to B) or with C (if you preferred B to D). This makes a total of four comparisons – two to put A, B, and C in order and two to find out where to put D.

We have not finished analysing the problem, because you may not have been lucky with A, B, and C. Perhaps all you know after the first two comparisons is that both A and C are preferable to B. Then you have another dilemma: is it better to compare A with C or to compare D with one of A, B, and C – and in the second case, should D be compared with B or with one of A and C? And once you have finished looking at those cases and subcases, you still have to see what would have happened if your second comparison had been between C and D.

The analysis becomes somewhat tedious, but it can be done. It shows that five comparisons are always enough, that sometimes you need that many, and that the second comparison should indeed be between C and D.

The trouble with this sort of argument is that the number of cases one must consider becomes very large very quickly. It would be out of the question to work out exactly how many comparisons are needed when you have, say, 100 objects – almost certainly this will never be known. (I well remember my shock when I first heard a mathematician declare that the exact value of a certain quantity would never be known. Now I am used to the fact that this is the rule rather than the exception. The quantity in question was the Ramsey number $R(5, 5)$, the smallest number n for which in any group of n people there must be five who all know each other or five who are all new to each other.) Instead, therefore, one tries to find upper and lower bounds. For this problem, an upper bound of c_n means a procedure for sorting n objects using no more than c_n comparisons, and a lower bound of b_n means a proof that, no matter how clever you are, b_n comparisons will sometimes be necessary. This is an example of a problem where the best-known upper and lower bounds are within a multiplicative factor of each other: it is known that, up to a multiplicative constant, the number of comparisons needed to sort n objects is $n \log n$.

One way to see why this is interesting is to try to devise a sorting procedure for yourself. An obvious method is to start by finding the object that comes top, set it aside, and then repeat. To find the best object, compare the first two, then compare the winner with the third, and the winner of that with the fourth, and so on. This way, it takes $n - 1$ comparisons to find the best, then $n - 2$ to find the next best, and so on, making a total of $(n - 1) + (n - 2) + (n - 3) + \ldots +$ comparisons in all, which works out at about $n^2/2$.

Natural though this method is, if you use it, then you end up comparing *every* pair of objects, so it is in fact as inefficient as possible (though it does have the advantage of being simple to program). When n is large, $n \log n$ is a very significant improvement on $n^2/2$, because $\log n$ is much smaller than $n/2$.

The following method, known as Quicksort, is not guaranteed to work any faster, but usually it works much faster. It is defined recursively (that is, in terms of itself) as follows. First choose any one of the objects, x, say, and arrange the others into two piles, the ones that are better than x and the ones that are worse. This needs $n - 1$ comparisons. All you need to do now is sort the two piles – which you do using Quicksort. That is, for each pile you choose one object and arrange the others into two further piles, and so on. Usually, unless you are unlucky, when you divide a pile into two further ones, they will be of roughly the same size. Then it can be shown that the number of comparisons you make will be roughly $n \log n$. In other words, generally this method works as well as you could possibly hope for, to within a multiplicative constant.

Chapter 8
Some frequently asked questions

1. Is it true that mathematicians are past it by the time they are 30?

This widely believed myth derives its appeal from a misconception about the nature of mathematical ability. People like to think of mathematicians as geniuses, and of genius itself as an utterly mysterious quality which a few are born with, and which nobody else has the slightest chance of acquiring.

The relationship between age and mathematical output varies widely from one person to another, and it is true that a few mathematicians do their best work in their 20s. The great majority, though, find that their knowledge and expertise develop steadily throughout their life, and that for many years this development more than compensates for any decline that there might be in 'raw' brain power – if that concept even makes sense. It is true that not many major breakthroughs are achieved by mathematicians over the age of 40, but this may well be for sociological reasons. By the age of 40, somebody who is capable of making such a breakthrough will probably already have become well known as a result of earlier work, and may therefore not have quite the hunger of a younger, less established mathematician. But there are many counter-examples to this, and some mathematicians continue, with enthusiasm undimmed, well past retirement.

In general, the popular view of a stereotypical mathematician – very clever perhaps, but also peculiar, badly dressed, asexual, semi-autistic – is not a flattering one. A few mathematicians do conform to the stereotype to some extent, but nothing would be more foolish than to think that if you do not, then you cannot be any good at mathematics. Indeed, all other things being equal, you may well be at an advantage. Only a very small proportion of mathematics students end up becoming research mathematicians. Most fall by the wayside at an earlier stage, for example by losing interest, not getting a PhD place, or doing a PhD but not getting a university job. It is my impression, and I am not alone in thinking this, that, among those who do survive the various culls, there is usually a smaller proportion of oddballs than in the initial student population.

While the negative portrayal of mathematicians may be damaging, by putting off people who would otherwise enjoy the subject and be good at it, the damage done by the word 'genius' is more insidious and possibly greater. Here is a rough and ready definition of a genius: somebody who can do easily, and at a young age, something that almost nobody else can do except after years of practice, if at all. The achievements of geniuses have a sort of magic quality about them – it is as if their brains work not just more efficiently than ours, but in a completely different way. Every year or two a mathematics undergraduate arrives at Cambridge who regularly manages to solve in a few minutes problems that take most people, including those who are supposed to be teaching them, several hours or more. When faced with such a person, all one can do is stand back and admire.

And yet, these extraordinary people are not always the most successful research mathematicians. If you want to solve a problem that other professional mathematicians have tried and failed to solve before you, then, of the many qualities you will need, genius as I have defined it is neither necessary nor sufficient. To illustrate with an extreme example, Andrew Wiles, who (at the age of just over 40) proved Fermat's Last Theorem (which states that if x, y, z, and n are all positive integers and n is greater than 2, then $x^n + y^n$ cannot equal z^n) and thereby solved the world's most famous unsolved mathematics problem, is undoubtedly very clever, but he is not a genius in my sense.

How, you might ask, could he possibly have done what he did without some sort of mysterious extra brainpower? The answer is that, remarkable though his achievement was, it is not so remarkable as to defy explanation. I do not know precisely what enabled him to succeed, but he would have needed great courage, determination, and patience, a wide knowledge of some very difficult work done by others, the good fortune to be in the right mathematical area at the right time, and an exceptional strategic ability.

This last quality is, ultimately, more important than freakish mental speed: the most profound contributions to mathematics are often made by tortoises rather than hares. As mathematicians develop, they learn various tricks of the trade, partly from the work of other mathematicians and partly as a result of many hours spent thinking about mathematics. What determines whether they can use their expertise to solve notorious problems is, in large measure, a matter of careful planning: attempting problems that are likely to be fruitful, knowing when to give up a line of thought (a difficult judgement to make), being able to sketch broad outlines of arguments before, just occasionally, managing to fill in the details. This demands a level of maturity, which is by no means incompatible with genius, but which does not always accompany it.

2. Why are there so few women mathematicians?

It is tempting to avoid this question, since giving an answer presents such a good opportunity to cause offence. However, the small proportion of women, even today, in the mathematics departments of the world, is so noticeable and so very much a fact of mathematical life, that I feel compelled to say something, even if what I say amounts to little more than that I find the situation puzzling and regrettable.

One point that deserves to be made is that the lack of women in mathematics is another statistical phenomenon: incredibly good female mathematicians exist and, just like their male counterparts, they have many different ways of being good, including, in some cases, being geniuses. There is no evidence whatsoever for any sort of upper limit on what women can achieve mathematically. Occasionally, one reads that men perform better at certain mental tests, of visuo-spatial ability, for example, and it is sometimes suggested that this accounts for their domination of mathematics. However, this sort of argument is not very convincing: visuo-spatial ability can be developed with practice, and, in any case, while it can sometimes be useful to a mathematician, it is rarely indispensable.

More plausible is the idea that social factors are important: where a boy may be proud of his mathematical ability, one can imagine a girl being embarrassed to excel at a pursuit that is perceived as unfeminine. In addition, mathematically gifted girls have few role models, so the situation is self-perpetuating. A social factor that may operate at a later stage is that mathematics, more than most academic disciplines, demands a certain single-mindedness which is hard, though certainly not impossible, to combine with motherhood. The novelist Candia McWilliam once said that each of her children had cost her two books; but at least it is possible to write a novel after a few years of not having done so. If you give up mathematics for a few years then you get out of the habit, and mathematical comebacks are rare.

It has been suggested that female mathematicians tend to develop later than their male counterparts and that this puts them at a disadvantage in a career structure that rewards early achievement. The life stories of many of the most prominent female mathematicians bear this out, though their late development is largely for the social reasons just mentioned, and again there are many exceptions.

None of these explanations seems sufficient, though. Rather than speculating any further, the best I can do is point out that several books have been written on this subject (see Further reading). A final comment is that the situation is improving: the proportion of women among mathematicians has steadily increased in recent years and, given how society in general has changed and is changing, it will almost certainly continue to do so.

3. Do mathematics and music go together?

Despite the fact that many mathematicians are completely unmusical and few musicians have any interest in mathematics, there is a persistent piece of folk knowledge that the two are connected. As a result, nobody is surprised to learn that a mathematician is a very good pianist, or composes music as a hobby, or loves to listen to Bach.

There is plenty of anecdotal evidence suggesting that mathematicians are drawn to music more than to any other art form, and some studies have claimed to demonstrate that children who are educated musically perform better in scientific subjects. It is not hard to guess why this might be. Although abstraction is important in all the arts, and music has a representational component, music is the most obviously abstract art: a large part of the pleasure of listening to music comes from a direct, if not wholly conscious, appreciation of pure patterns with no intrinsic meaning.

Unfortunately, the anecdotal evidence is backed up by very little hard science. It is not even obvious what questions should be asked. What would we learn if statistically significant data were collected showing that a higher percentage of mathematicians played the piano than of non-mathematicians with similar social and educational backgrounds? My guess is that such data *could* be collected, but it would be far more interesting to produce an experimentally testable theory that explained the connection. As for statistical evidence, this would be much more valuable if it was more specific. Both mathematics and music are very varied: it is possible to be passionately enthusiastic about some parts and completely uninterested in others. Are there subtle relationships between mathematical and musical tastes? If so, then they would be much more informative than crude correlations between interest in the disciplines as a whole.

4. Why do so many people positively dislike mathematics?

One does not often hear people saying that they have never liked biology, or English literature. To be sure, not everybody is excited by these subjects, but those who are not tend to understand perfectly well that others are. By contrast, mathematics, and subjects with a high mathematical content such as physics, seem to provoke not just indifference but actual antipathy. What is it that causes many people to give mathematical subjects up as soon as they possibly can and remember them with dread for the rest of their lives?

Probably it is not so much mathematics itself that people find unappealing as the experience of mathematics lessons, and this is easier to understand. Because mathematics continually builds on itself, it is important to keep up when learning it. For example, if you are not reasonably adept at multiplying two-digit numbers together, then you probably won't have a good intuitive feel for the distributive law (discussed in Chapter 2). Without this, you are unlikely to be comfortable with multiplying out the brackets in an expression such as $(x + 2)(x + 3)$, and then you will not be able to understand quadratic equations properly. And if you do not understand quadratic equations, then you will not understand why the golden ratio is $\dfrac{1 + \sqrt{5}}{2}$.

There are many chains of this kind, but there is more to keeping up with mathematics than just maintaining technical fluency. Every so often, a new idea is introduced which is very important and markedly more sophisticated than those that have come before, and each one provides an opportunity to fall behind. An obvious example is the use of letters to stand for numbers, which many find confusing but which is fundamental to all mathematics above a certain level. Other examples are negative numbers, complex numbers, trigonometry, raising to powers, logarithms, and the beginnings of calculus. Those who are not ready to make the necessary conceptual leap when they meet one of these ideas will feel insecure about all the mathematics that builds on it. Gradually they will get used to only half understanding what their mathematics teachers say, and after a few more missed leaps they will find that even half is an overestimate. Meanwhile, they will see others in their class who are keeping up with no difficulty at all. It is no wonder that mathematics lessons become, for many people, something of an ordeal.

Is this a necessary state of affairs? Are some people just doomed to dislike mathematics at school? Or might it be possible to teach the subject differently in such a way that far fewer people are excluded from it? I am convinced that any child who is given one-to-one tuition in mathematics from an early age by a good and enthusiastic teacher will grow up liking it. This, of course, does not immediately suggest a feasible educational policy, but it does at least indicate that there might be room for improvement in how mathematics is taught.

One recommendation follows from the ideas I have emphasized in this book. Above, I implicitly drew a contrast between being technically fluent and understanding difficult concepts, but it seems that almost everybody who is good at one is good at the other. And indeed, if understanding a mathematical object is largely a question of learning the rules it obeys rather than grasping its essence, then this is exactly what one would expect – the distinction between technical fluency and mathematical understanding is less clear-cut than one might imagine.

How should this observation influence classroom practice? I do not advocate any revolutionary change – mathematics has suffered from too many of them already – but a small change in emphasis could pay dividends. For example, suppose that a pupil makes the common mistake of thinking that $x^{a+b} = x^a + x^b$. A teacher who has emphasized the intrinsic meaning of expressions such as x^a will point out that x^{a+b} means $a + b$ xs all multiplied together, which is clearly the same as a of them multiplied together *multiplied* by b of them multiplied together. Unfortunately, many children find this argument too complicated to take in, and anyhow it ceases to be valid if a and b are not positive integers.

Such children might benefit from a more abstract approach. As I pointed out in Chapter 2, everything one needs to know about powers can be deduced from a few very simple rules, of which the most important is $x^{a+b} = x^a x^b$. If this rule has been emphasized, then not only is the above mistake less likely in the first place, but it is also easier to correct: those who make the mistake can simply be told that they have forgotten to apply the right rule. Of course, it is important to be familiar with basic facts such as that x^3 means x times x times x, but these can be presented as consequences of the rules rather than as justifications for them.

I do not wish to suggest that one should try to explain to children what the abstract approach is, but merely that teachers should be aware of its implications. The main one is that it is quite possible to learn to use mathematical concepts correctly without being able to say exactly what they mean. This might sound a bad idea, but the use is often easier to teach, and a deeper understanding of the meaning, if there *is* any meaning over and above the use, often follows of its own accord.

5. Do mathematicians use computers in their work?

The short answer is that most do not, or at least not in a fundamental way. Of course, just like anybody else, we find them indispensable for word-processing and for communicating with each other, and the Internet is becoming more and more useful. There are some areas of mathematics where long, unpleasant but basically routine calculations have to be done, and there are very good symbolic manipulation programs for doing them.

Thus, computers can be very useful time-saving devices, sometimes so much so that they enable mathematicians to discover results that they could not have discovered on their own. Nevertheless, the kind of help that computers can provide is very limited. If it happens that your problem, or more usually sub-problem, is one of the small minority that can be solved by a long and repetitive search, then well and good. If, on the other hand, you are stuck and need a bright idea, then, in the present state of technology, a computer will be no help whatsoever. In fact, most mathematicians would say that their most important tools are a piece of paper and something to write with.

My own view, which is a minority one, is that this is a temporary state of affairs, and that, over the next hundred years or so, computers will gradually be able to do more and more of what mathematicians do – starting, perhaps, with doing simple exercises for us or saving us from wasting a week trying to prove a lemma to which a well-known construction provides a counter-example (here I speak from frequent experience) and eventually supplanting us entirely. Most mathematicians are far more pessimistic (or should that be optimistic?) about how good computers will ever be at mathematics.

6. How is research in mathematics possible?

Conversely, one might ask, what is it that seems so paradoxical about the possibility of mathematical research? I have mentioned several unsolved problems in this book, and mathematical research consists, in large measure, in trying to solve those and similar ones. If you have read Chapter 7, then you will see that a good way to generate questions is to take a mathematical phenomenon that is too hard to analyse exactly, and try to make approximate statements about it. Another method is suggested by the end of Chapter 6: choose a difficult mathematical concept, such as a four-dimensional manifold, and you will usually find that even simple questions about it can be very hard to answer.

If there is a mystery about mathematical research, it is not that hard questions exist – it is in fact quite easy to invent impossibly hard questions – but rather that there are enough questions of just the right level of difficulty to keep thousands of mathematicians hooked. To do this, they must certainly be challenging but they must also offer a glimmer of hope that they can be solved.

7. Are famous mathematical problems ever solved by amateurs?

The simplest and least misleading answer to this question is a straightforward no. Professional mathematicians very soon learn that almost any idea they have about any well-known problem has been had by many people before them. For an idea to have a chance of being new, it must have some feature that explains why nobody has previously thought of it. It may be simply that the idea is strikingly original and unexpected, but this is very rare: on the whole, if an idea comes, it comes for a good reason rather than simply bubbling up out of nowhere. And if it has occurred to you, then why should it not have occurred to somebody else? A more plausible reason is that it is related to other ideas which are not particularly well known but which you have taken the trouble to learn and digest. That at least reduces the probability that others have had it before you, though not to zero.

Mathematics departments around the world regularly receive letters from people who claim to have solved famous problems, and virtually without exception these 'solutions' are not merely wrong, but laughably so. Some, while not exactly mistaken, are so unlike a correct proof of anything that they are not really attempted solutions at all. Those that follow at least some of the normal conventions of mathematical presentation use very elementary arguments that would, had they been correct, have been discovered centuries ago. The people who write these letters have no conception of how difficult mathematical research is, of the years of effort needed to develop enough knowledge and expertise to do significant original work, or of the extent to which mathematics is a collective activity.

By this last point I do not mean that mathematicians work in large groups, though many research papers have two or three authors. Rather, I mean that, as mathematics develops, new techniques are invented that become indispensable for answering certain kinds of questions. As a result, each generation of mathematicians stands on the shoulders of previous ones, solving problems that would once have been regarded as out of reach. If you try to work in isolation from the mathematical mainstream, then you will have to work out these techniques for yourself, and that puts you at a crippling disadvantage.

This is not quite to say that no amateur could ever do significant research in mathematics. Indeed, there are one or two examples. In 1975 Marjorie Rice, a San Diego housewife with very little mathematical training, discovered three previously unknown ways of tiling the plane with (irregular) pentagons after reading of the problem in the *Scientific American*. And in 1952 Kurt Heegner, a 58-year-old German schoolmaster, proved a famous conjecture of Gauss which had been open for over a century.

However, these examples do not contradict what I have been saying. There are some problems which do not seem to relate closely to the main body of mathematics, and for those it is not particularly helpful to know existing mathematical techniques. The problem of finding new pentagonal tilings was of such a kind: a professional mathematician would not have been much better equipped to solve it than a gifted amateur. Rice's achievement was rather like that of an amateur astronomer who discovers a new comet – the resulting fame is a well-deserved reward for a long search. As for Heegner, though he was not a professional mathematician, he certainly did not work in isolation. In particular, he had taught himself about modular functions. I cannot explain what these are here – indeed, they would normally be considered too advanced even for an undergraduate mathematics course.

Interestingly, Heegner did not write up his proof in a completely conventional way, and although his paper was grudgingly published it was thought for many years to be wrong. In the late 1960s, the problem was solved again, independently, by Alan Baker and Harold Stark, and only then was Heegner's work carefully re-examined and found to be correct after all. Unfortunately, Heegner died in 1965 and thus did not live to see his rehabilitation.

8. Why do mathematicians refer to some theorems and proofs as beautiful?

I have discussed this question earlier in the book, so here I will be very brief. It may seem odd to use aesthetic language about something as apparently dry as mathematics, but, as I explained in Chapter 2 (at the end of the discussion of the tiling problem), mathematical arguments can give pleasure, and this pleasure has many features in common with more conventional aesthetic pleasure.

One difference is that, at least from the aesthetic point of view, a mathematician is more anonymous than an artist. While we may greatly admire a mathematician who discovers a beautiful proof, the human story behind the discovery eventually fades away and it is, in the end, the mathematics itself that delights us.

Further reading

There are some important aspects of mathematics that I have not had the space to discuss here. For these I can recommend other books. If you want to learn about the history of mathematics, it is hard to beat Morris Kline's magisterial three volumes on the subject, *Mathematical Thought from Ancient to Modern Times* (Oxford University Press, 1972), though he expects more mathematical sophistication from his readers than I have here. *Innumeracy*, by John Allen Paulos, (Viking, 1989) has rapidly become a classic on the subject of how knowledge of mathematics can influence one's judgements in everyday life—for the better. Tom Körner's *The Pleasures of Counting* (Cambridge University Press, 1996) says much more about the applications of mathematics than I have, and does so more wittily. *What is Mathematics?* by Courant and Robbins (Oxford University Press, 2nd edn., 1996) is another classic. It is similar in spirit to this book, but longer and somewhat more formal. *The Mathematical Experience* by Davis and Hersch (Birkhäuser, 1980) is a delightful collection of essays about mathematics, written in a philosophical vein. I would have liked to say more about probability, but a beautiful discussion of randomness and its philosophical implications can instead be found in *Mathematics and the Unexpected*, by Ivar Ekeland (University of Chicago Press, 1988).

The quotations on page 18 are from Saussure's *Course in General Linguistics* (McGraw-Hill, 1959) and Wittgenstein's *Philosophical Investigations* (Blackwell, 3rd edn., 2001). Anybody who has read this book and the *Philosophical Investigations* will see how much the later Wittgenstein has influenced my philosophical outlook and in particular my views on the abstract method. Russell and Whitehead's famous *Principia Mathematica* (Cambridge University Press, 2nd edn., 1973) is not exactly light reading, but if you found some of my proofs of elementary facts long-winded, then for comparison you should look up their proof that $1 + 1 = 2$. On the subject of women in mathematics, discussed in Chapter 8, two good recent books are *Women in Mathematics: The Addition of Difference* by Claudia Henrion (Indiana University Press, 1997) and *Women Becoming Mathematicians: Creating a Professional Identity in Post-World War II America* by Margaret Murray (MIT Press, 2000).

Finally, if you have enjoyed this book, you might like to know that in order to keep it very short I reluctantly removed whole sections, including a complete chapter, from earlier drafts. Some of this material can be found on my home page:

http://www.dpmms.cam.ac.uk/~wtg10

Index

Mathematics

Expand your collection of
VERY SHORT INTRODUCTIONS

Available now

Visit the
VERY SHORT INTRODUCTIONS
Web site

www.oup.co.uk/vsi

➤ **Information** about all published titles

➤ News of **forthcoming books**

➤ **Extracts** from the books, including titles not yet published

➤ **Reviews** and views

➤ **Links** to other **web sites** and main OUP web page

➤ Information about **VSIs in translation**

➤ **Contact** the editors

➤ **Order** other **VSIs** on-line